Introduction to Molecular Genomics

Authored by

Maryam Javed

Institute of Biochemistry and Biotechnology
University of Veterinary and Animal Sciences
Lahore
Pakistan

Asif Nadeem

Department of Biotechnology
Virtual University of Pakistan
Lahore
Pakistan

&

Faiz-ul Hassan

Institute of Animal & Dairy Sciences
University of Agriculture
Faisalabad
Pakistan

Introduction to Molecular Genomics

Authors: Maryam Javed, Asif Nadeem and Faiz-ul Hassan

ISBN (Online): 978-1-68108-925-6

ISBN (Print): 978-1-68108-926-3

ISBN (Paperback): 978-1-68108-927-0

need for a court order if at any point you breach any terms of this License Agreement. In no event will any delay or failure by Bentham Science Publishers in enforcing your compliance with this License Agreement constitute a waiver of any of its rights.

3. You acknowledge that you have read this License Agreement, and agree to be bound by its terms and conditions. To the extent that any other terms and conditions presented on any website of Bentham Science Publishers conflict with, or are inconsistent with, the terms and conditions set out in this License Agreement, you acknowledge that the terms and conditions set out in this License Agreement shall prevail.

Bentham Science Publishers Ltd.
Executive Suite Y - 2
PO Box 7917, Saif Zone
Sharjah, U.A.E.
Email: subscriptions@benthamscience.net

BENTHAM SCIENCE

DEDICATION

Dedicated to

James Watson and Francis Crick
For unwinding the complex helical of life to explore and conquer

CONTENTS

FOREWORD

Molecular genomics is currently navigating through a revolutionary era promoted by advances in DNA technologies. Knowledge web in this field has been developed with an extraordinary potential over the last century. The innovative work in this field has changed how we view and understand biological systems and this new knowledge has led the way to innovative opportunities, allowing the subject to grow and evolve at a phenomenal rate. Molecular biology underpins much of today's biological research. Molecular biology is the study of the structure and function of biologically important molecules, including DNA, RNA, proteins, and the molecular events that govern cell function. Molecular biology techniques such as molecular cloning of genes, gene transfer, genetic manipulation of animal and plant embryo transfer, genetic manipulation of rumen microbes, chemical and biological treatment of low-quality animal feed for improved nutritive value, genetically engineered immunodiagnostic and immunoprophylactic agents, as well as veterinary vaccines, are a reality today and are finding their ways into research and development program of global significance.

The structure and literary flavor of the book is up to the expectations of the academicians and scholars. This book covers significant topics of the subject like genomics and its spectrum, modern techniques, an overview of genetic engineering, pan genomics, the role of genomics in the environment, the significance of digital genomic era and the contribution of genomics towards the society. Although the spectrum of the subject is immense, I am glad to see that the book covers all relevant areas which are essential to understand the basic concepts of the subject.

In today's world of unprecedented knowledge and information, such a fine book is worth reading. This book is highly recommended, especially for graduate students who are in their early knowledge capacity. I would like to congratulate the authors for contributing in bridging the knowledge gap of basic principles of molecular genomics. I hope this book will be a fine addition to the field of genomics and basic molecular biology.

Prof. Dr. İbrahim Hakkı CİĞERCİ
Department of Molecular Biology and Genetics
Afyonkocatepe University Afyonkarahisar
Turkey

PREFACE

Molecular biology is currently navigating through a genomic era promoted by advances in DNA technologies. New advances continue, and efforts are currently placed in whole genome sequencing for its future implementation to improve the accuracy of genomic selection or mapping new QTL of interest. Most modern sciences are well understood by scholars at the postgraduate level but there are very limited books and literature available targeting the undergraduate students. There is a need to develop literature, which is easily understandable for our young undergraduates, so they can also get knowledge of various molecular techniques.

With this perspective in focus, now is the time to compile all the basic information of molecular genomics into a single portfolio, so that this information can be disseminated to its real beneficiaries (young scholars). This book comprises of major features as a historical perspective of genomics, with particular emphasis on animal genomes and their time-line in modern genomics, a major phenomenon in the molecular environment of the cell, evolutionary journey of a single cell into multiplex and advanced form of life, comparison of expression patterns in different cellular types, the importance of species diversity and its molecular basis, the impact of biodiversity on socio-economical strength of a nation, conservation efforts at the genomic level for a sustainable ecosystem. After this basic introduction to genomics, the book highlights the State-Of-The-Art techniques used in the modern era of animal genomics for the advancement of economically significant traits. Gene manipulation and genetic engineering to develop better animal species have also been included. The effect of genomic environment on the expression of genes is also an essential part of defining the phenotype of a specie. Finally, ethical and moral aspects of animal genomics have been discussed to shape the Dos and DON'Ts in modern sciences. All the contents included in the book have been designed in accordance with the needs of undergraduate students. In the past few decades, there has been tremendous expansion in the field of genomics and many innovative modern methodologies have been reported that are equally useful in human and animal welfare. But most written content, in terms of books, papers, monographs and articles, focus on the worth of these scientific interventions in human sciences and less literature can be found on the emphasis of these advancements in animal sciences. In developing countries like Pakistan, where students assess modern advances in a limit, there is a need to establish a national information hub in the form of written materials that can be recommended at undergraduate and post graduate levels to uplift our scholar's intellectual abilities by giving such exposure of modern advances. This book is a humble effort to contribute towards the rational capacity building of our young scholars, so they can be skilled in accordance with leading scientific advances in molecular genomics.

CONSENT FOR PUBLICATION

Not applicable.

CONFLICT OF INTEREST

The authors declare no conflict of interest, financial or otherwise.

ACKNOWLEDGEMENTS

Declared none.

Maryam Javed
Institute of Biochemistry and Biotechnology
University of Veterinary and Animal Sciences
Lahore
Pakistan

Asif Nadeem
Department of Biotechnology
Virtual Uuniversity of Pakistan
Lahore
Pakistan

&

Faiz-ul Hassan
Institute of Animal & Dairy Sciences
University of Agriculture
Faisalabad
Pakistan

CHAPTER 1

Genome: A Code of Life

Abstract: This chapter proposes a brief description of deoxyribonucleic acid: Molecule of life eventually decode; Molecular organization of Genome; Evolutionary journey of a single cell to a multiplex form of life; Genome Replication; deoxyribonucleic acid Damage, Mutations and Repair.

Keywords: Damage, Genome, Mutation, Repair and Replication.

The fascinating and complex world of genomics is all about the phenotypic manifestation of genetic information into the complex traits controlling all biological events in a cell and ultimately in an individual. From a historical perspective, genomics is a relatively new science. Around the turn of the twentieth century, major principles of the centralized controlling system of biology were discovered and rationalized to dig into the philosophy of life at micro and molecular levels. Major happenings of that era helped define the current face of advanced genomics with lots of applications in medicine, industry, agriculture, food, environment, *etc*. Let's have a flashback on the major events in the historical timeline of genomics.

DNA: MOLECULE OF LIFE EVENTUALLY DECODE

Deoxyribonucleic acid was first brought into light by a 24-year-old Swiss physician and biochemist, *Johann Friedrich Miescher,* in 1868 during his stay in Felix Hoppe-Seyler laboratory in Tübingen University. He isolated DNA from the nucleus of white blood cells, but he never determined its purpose or function until 1943. Before that time, it was widely believed that proteins store genetic information. In his experiments, Miescher characterized the properties of this molecule and noticed that this molecule was neither a lipid nor a protein. Analysis of its structure disclosed that it has sulphur. Later, Miescher confirmed that sulphur is not present, but phosphorous is present in a large amount. This was firstly named Nuclein (T. F. Lee, 2013).

In his further investigation, Miescher also postulated that this molecule was an integral part of all nuclei and could have a role in the functionality of the cells. Miescher hypothesized that it might have a part in transmitting traits to the next generation, but this idea was rejected later on. Later, Oswald Avery, Maclyn McCarty, and colleagues studied the phenomenon of transformation in bacteria and reported that DNA was the material transferred across bacterial cells restoring the pathogenicity of Pneumococcal bacterial strains. Positive responses were received about this idea, and work was appreciated. But regardless of this appreciation, Avery's work was rejected by the scientific community of the Nobel Foundation. Later, Nobel Foundation apologized to the public for failing to award Avery a Nobel Prize (Dahm, 2005).

In 1952, further investigation on DNA as a speculated genetic material was reported by Chase and Hershey, which aided in verifying that DNA is the genetic material. Chase and Hershey confirmed that the DNA from the phage is injected into the bacteria after the virus is attached to the host bacterium. They used labeled ^{32}P in DNA strands, which were detected inside the bacterial cell, while protein coats coated with ^{35}S remained outside the cell (Barbieri, 2007);(Fridell, 2005).

Meanwhile, the chemical nature of DNA molecules was also rigorously studied. It was found that this molecule was composed of sugar-phosphate and nitrogen-containing four different types of bases: Adenine, Guanine, Cytosine, and Thymine. Austrian chemist Erwin Chargaff reported another historical finding in 1952 regarding the base pairing of the bases. According to Chargaff's rules, nitrogenous bases adenine, guanine, thymine and cytosine have some ratio in DNA strands. Guanine is always equal in number with cytosine and adenine is always equal in number with thymine. In other words, someone can say that DNA from any cell should have a 1:1 ratio of purine and pyrimidine bases according to Base Pair Rule. This discovery provided a baseline for the dramatic breakthrough of the century:

Discovery of the Structure of DNA Molecule

DNA was initially structured as a triple helix molecule by Pauling in 1953, but he was not alone in the quest of DNA structure. James Watson, a biologist, Francis Crick (Physicist), Wilkins and Franklin were also trying hard to put the evidence together for the missing puzzle of DNA structure (As given in the famous Photo-51) (Maddox, 2002). This crystallographic view indicates that DNA was crystallized and appeared as fuzzy X, showing a helical structure (Fig **1.1**). When Watson collected a copy of this image, he saw the significance of the X. Crick, and Watson demonstrated the DNA as a double helical structure, with anti-parallel strands with nitrogenous bases for stabilizing molecules (Jd & Fh, 1953).

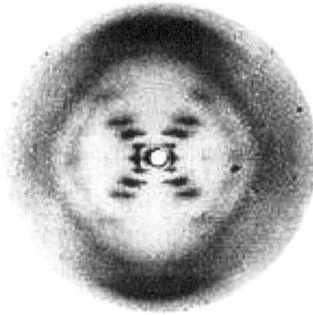

Fig. (1.1). X-Ray Crystallographic view of DNA.

Molecular Organization of Genome

An organism's genome is its complete set of DNA which is also a source of transmitting genetic information through generations. With its four-letter language, DNA is a code for all phenotypic characteristics in any organism. A gene is defined as a hereditary unit of DNA that transmits commands for synthesizing proteins of different functions. Approximately 3 billion DNA base pairs are present in each human body cell that makes the human genome. Typically, DNA is a double-helical assembly and has two strands running in opposite directions. (There are some examples of viral DNA which are single-stranded). Nucleotides are present in the DNA. Nucleotides are polymers of subunits named polynucleotides (Wachsmuth, Caudron-Herger, & Rippe, 2008);(Bernardi, 1995). DNA strands have sugar molecules that are linked with phosphate groups. These molecules make the backbone of DNA strands. There are five ringed carbons present in sugar molecules. The carbon number 3 of ribose sugar is bonded with the phosphate group of carbon number 5 of the next deoxyribose sugar. Hence the connection is known as 3' to 5' phosphodiester bond. Both DNA strands are demonstrated as 5' to the 3' end, where 5' ends in a phosphate group and 3' ends in a sugar molecule. The deoxyribose sugar is covalently linked to one of the four nitrogenous bases (Guanine, Thymine, Cytosine and Adenine). A and G are collectively called purines and they are double-ringed molecules, while C and T are collectively called pyrimidines and are single-ringed molecules (Rippe, 2012).

Double-stranded DNA strands run in the opposite direction. These two strands are coiled together by H-bonding in a particular manner, where A develops a double electron pair sharing with T and C depicts a triple H-bond with G. The G-C interaction carries more bond energy and is almost 30 percent stronger than A-T interaction. DNA segments with more A-T rich bases are more liable to thermal instabilities due to less break bond energy. These nitrogenous bases are arranged upright to the central axis. Being hydrophobic in nature, these bases could not develop hydrogen bonds with water molecules. The interaction energy is a combination of hydrogen bonding between complementary nitrogenous bases. Even in the single-stranded state, the bases favor to be piled, and a single-stranded chain can also have regions of helical conformation.

The backbone in DNA has a negative charge. In the absence of a salty surrounding medium, these charges cause a strong repulsive force between the double helix of DNA and keep them separated from each other. Therefore, there is a need for counter-ions to maintain the helical arrangement intact. These ions protect the charges on the backbone and give a good force, as analogous to the Van Der Waals force for fluctuating induced dipoles (Sinden, 1994).

> The first person to give any thought to the three dimensional structure of DNA was **W. T. Astbury** who by his X-ray crystallographic studies of DNA molecule concluded in 1940's that because DNA has high density, so, its polynucleotide was a stack of flat nucleotides, each of which was oriented perpendicularly to the long axis of molecule and was situated every 3.4 A along the stack.

Table 1. Types of DNA with differences of structural configuration.

	A-DNA Right Handed Helix Broad and Short	**B-DNA** Right-handed Thin and Long	**Z-DNA** Left Handed Thin and longer
Helix Diameter	25.5^0A	23.7^0A	18.4^0A
Bp	2.3^0A	3.4^0A	3.8^0A
Bp/Helical Turn	11	10	12
Helix Pitch	25^0A	34^0A	47^0A
Twist of the base	20deg	-1deg	-9deg

B-DNA is the most commonly famous form of DNA. Under low hydration conditions and applied force or twist conditions, DNA can adopt many other configurations, like Z-DNA, S-DNA, and A-DNA *etc.* (Table **1**).

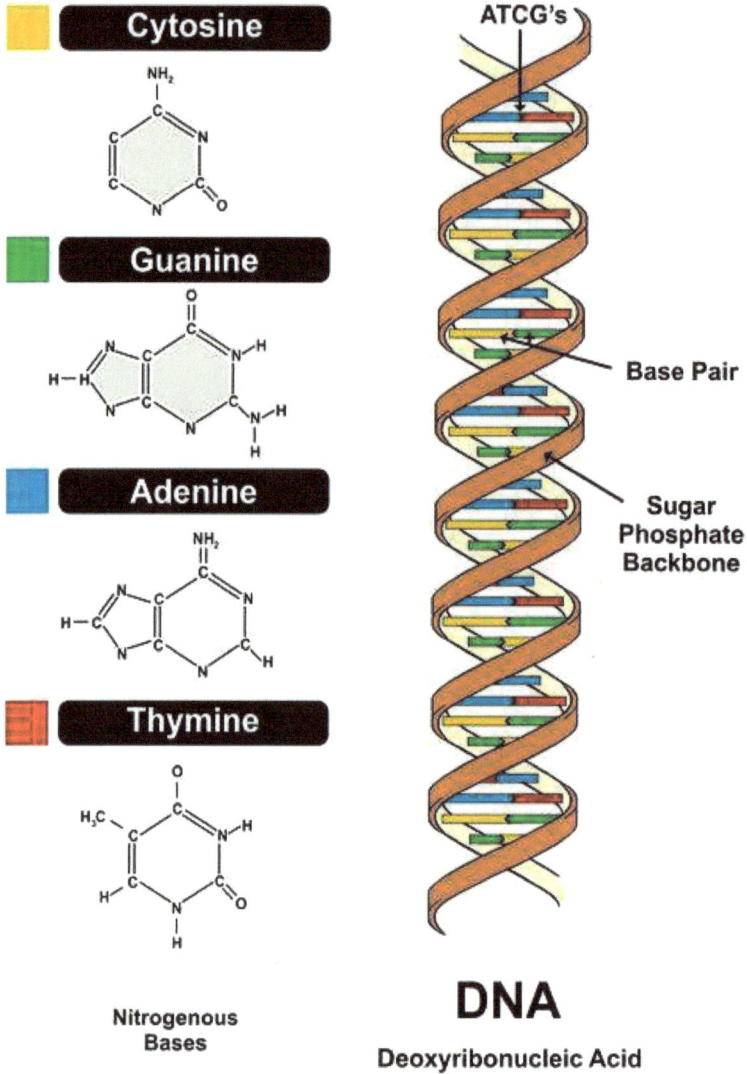

Fig. (1.2). Structure of DNA molecule. (www.livescience.com).

If the entire DNA in your body was put
end to end, it would reach to the sun
and back over 600 times.

During further structural investigations of DNA molecules, another double helix was found. DNA antiparallel strands are not located evenly with respect to each other; grooves are unequal in size and are termed as major and minor grooves (Fig. **1.2**). This variation is due to the unequal proportion of carbon atoms in the four bases. Two of them (A/G) carry a double-ringed structure called purines, while Cytosine and Thymine are a single ringed structure known as pyrimidines. The diameter of the Major groove is 22Å, and that of the minor groove is 12 Å. The size of the major groove means that the edges of nitrogenous bases are more nearby in the major groove than in that of the minor groove. These edges provide excellent places for the attachment of many regulatory proteins for transcription factors or enzymes for DNA replication that can bind to exact sequences in DNA which usually make contact with the sides of the bases present in the major groove. This situation differs in the uncommon structure of DNA, but minor and major grooves are named to reveal the differences in size at all times that can be seen if the DNA strands are twisted back into the B form (Hartl, 2014).

The packing of a molecule of DNA in smaller spaces of the nucleus (almost 2×10^{-5}-times smaller in length) is phenomenal and efficient. It is due to the attachment of a special class of proteins that condense it in the form of Chromatin. These proteins, called Histones, form a complex molecule after organizing DNA at the first level called a nucleosome. This is made of two copies of each of histone proteins H2A, H2B, H3, and H4, accumulated in an octameric core with DNA wrapped firmly around it. 146-147 bp of DNA is used with histone proteins (Ramani, Shendure, & Duan, 2016).

Evolutionary Journey of a Single Cell to Multiplex form of Life

Living things have developed into three major groups of closely related beings, called "domains": Eukaryota, Archaea, and Bacteria. Fossil records specify that almost 2 billion years ago, eukaryotic cells developed from prokaryotic cells. How the first cell originated and how life came into being are matters of assumption since these happenings cannot be imitated in the laboratory. However, many laboratory trials give important indications bearing on some stages of the process. In the 1920s, it was first postulated that simple organic molecules could be converted into complex macromolecules through spontaneous polymerization. But for this, we need to develop the same climatic conditions prevailing on earth when it initially started. At the time of life initiation on earth, the earth's atmosphere was filled with CO_2, and N_2 gasses, and there was no or very little free Oxygen available. Along with these, some more gasses were also present in smaller volumes as H_2S, CO and H_2. This atmosphere supported the synthesis of different organic macromolecules in the presence of sunlight and electrical discharge. In the 1950s, the impulsive development of organic molecules was revealed firstly with the help of an experiment, when Stanley Miller displayed that the release of electric sparks into a mixture of CH_4, NH_3, and H_2, in the presence of water, directed to the development of a diversity of organic molecules, including numerous amino acids. Miller's experiments did not exactly mimic the conditions of primitive Earth; they obviously revealed the credibility of the spontaneous synthesis of organic molecules, given that the basic materials from which life arose.

Under plausible prebiotic conditions, the following stage in evolution was developing monomeric building blocks of macromolecules to polymerize suddenly. In hot climatic conditions, amino acid mixtures were converted into longer polypeptide chains. But the serious characteristic of the macromolecules must have been the capacity to duplicate itself from which life originated. Only a macromolecule adept at directing the formation of new copies of itself would have been competent of reproduction.

In eukaryotic cells, nucleic acids and proteins are the two main classes of imperative macromolecules. Out of these two molecules, only the nucleic acids can self-replicate. Each parent strand serves as a template for replication of other daughter DNA strands, which can be passed on to generations after generations. The discovery of RNA was another critical step in our understanding of the phenomenon of molecular evolution. In the 1980s, in the lab of Tom Cech and Sid Altman, it was found that RNA molecules have specific characteristics for catalyzing many chemical reactions, including nucleotides polymerization. In this way, RNA is exclusively able to assist as a template and as well as catalyze its own replication. Subsequently, RNA is commonly supposed to have been the most primeval genetic system, and the self-replicating RNA molecules phenomenon is believed to be an early stage of chemical evolution. This period of evolution is called as RNA world.

Even if we compare a bacterial cell with a human cell type, then we can find drastic changes in terms of complexity and quantity of biological entities. Many bacteria are rod-shaped, spiral, or spherical. Their diameters are from 1 to 10 micrometers. And almost 0.6 million to 5 million bp are present in their DNA contents; 5000 different proteins can be encoded with these base pairs. This DNA molecule is compacted into a single chromosome, but now we have several chromosomes carrying different genes in eukaryotes. This increase in the genomic content can be justified by the phenomenon of gene/chromosomal duplication at some specific hot spots called transposable elements, which are sites for endonucleases. This process will be elaborated in upcoming chapters. The first cell is supposed to have originated by the enclosure of RNA in a membrane that is made up of phospholipids.

In recent days, prokaryotic organisms like bacteria can be categorized into two major groups-eubacteria and archaebacterial-separated earlier in evolution. Archaebacteria can survive in harsh environmental conditions that are not present today but may be predominant in primeval Earth. Cyanobacteria and bacteria are the biggest and most complex prokaryotes, in which photosynthesis developed.

Among different prokaryotes, *Escherichia coli* are the most characterized organisms present in the human intestinal tract as normal flora. *E.coli* is bacillus bacteria. The diameter of this bacterium is 1 μm, and its length is almost 2 μm. Like all other bacteria, *Escherichia coli* are protected by a firm outer shell called cell wall, made up of peptides and polysaccharides. Plasmalemma is present within the cell wall. Plasmalemma is a typical lipid bilayer with embedded proteins. Being porous in nature, the cell wall can readily be pierced by a number of molecules; the cell membrane gives the selective and functional barrier between the cell and its surrounding environment. *E. coli* DNA has a single circular nature, while the eukaryotic nucleus has a membrane that separates it from the cell cytoplasm. Another significant structure in the cytoplasm is ribosomes, which are machinery for protein synthesis. These are about 30,000 in number and provide a granular appearance to the cell.

We see many organisms in our daily lives, like plants and animals,that have their place to the third domain (Eukaryote). Eukaryotes are different from prokaryotes in many aspects, as Eukaryotic DNA is linear and is enclosed inside a nucleus. The powerhouse is present in Eukaryotic cells, called mitochondria, which have a major role in producing energy for different cell activities and are crucial in the evolution.

These organelles are unique from other cellular components as they carry a self-replicating circular DNA molecule that is more likely to be a form of primitive genetic material, somehow managed to survive in advanced cellular forms. The complex form of life as eukaryotes started a whole new era of life on earth, by evolving into multicellular organisms. The invasion of prokaryotic cells is described below, which is based upon two proposed pathways (Barbieri, 2007).

Endosymbiosis

Margulis presented this hypothesis at Massachusetts University. He speculated that the two individual and independent prokaryotic cells merged together for mutual benefit or were engulfed by the other one and started living together. This theory explained the mitochondrion in animal cells and clarified the presence of chloroplast in plant cells. Mitochondria and chloroplasts both have their own DNA (which is circular) and RNA (O'Malley, 2015). This is the evidence of the Modern-day hypothesis supporting endosymbiosis (Fig. **1.4**). Both chloroplast and mitochondria reproduce by binary fission independent of the host cell metabolism. Thus, both appear to be more alike to prokaryotic cells than eukaryotic cells (Margulis & Chapman, 1998).

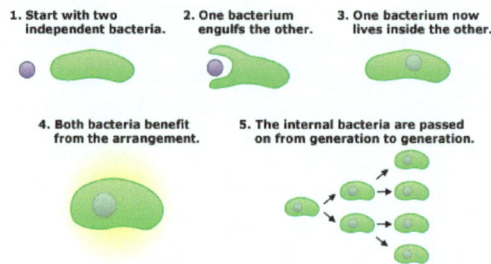

Fig. (1.3). Endosymbiosis in prokaryotic cells resulted in more complex for of life.

Membrane Infolding

Invasion of prokaryotic cells may be successful because the host cell membrane is enfolded to cover the invading prokaryotic cells, thus helping to transport them in cells. The membrane does not dissolve but remains intact, and thus creates a second membrane around the protochloroplast and protomitochondria. Present days show that the inner membrane of both the chloroplast and mitochondria contain more similar structures to prokaryotes than that of eukaryotic cells, whereas the outer membrane preserves eukaryote characteristics. It is also proposed that continuous membrane infolding formed the endomembrane system. It can be reported that possibly the first eukaryotic cell type was amazingly born from prokaryotic multicellular interactions (Margulis & Fester, 1991).

Genome Replication

Before the actual division, the cell must replicate its genome with high precision and accuracy (Müller *et al.*, 2014). This phenomenon of copying and transferring exact genomic information into daughter cells helps maintain genomic integrity and conserves the essential features of the genome. Genome replication is governed by many of the chemical players that maintain the flow of this process till both strands find their complimentary copies in a semi-conservation manner (Hand, 1978). Major steps involved in this process are as given:

DNA Templating

In DNA replication, parent nucleotide strands of DNA (A, T, C, and G) act as blueprints or templates to shape the new daughter strands. Free and unpolarized nucleotide bases identify their complementary partner to get attached and the newly formed strand elongates till the whole strand gets its complementary strand. According to the semi-conservation model of DNA replication, each DNA strand is used as a template for complementary DNA strand formation. Thus, original strands remain integral through many of the cell generations (O'Reilly, Turberfield, & Wilks, 2017).

Replication Fork

During the early 1960s, it was revealed that DNA replication is initiated on some hotspots instead of anywhere on a replicating chromosome. There was a discovery of localized regions of replication that move gradually together with the parental DNA strands. The replication fork is the active region and has a Y-shaped structure. This structure provides a place of attachment for many enzymes and proteins required to make a duplicated copy of the parent strand (Waga & Stillman, 1998).

DNA Unwinding and Topoisomerase

Being a supercoiled structure, the first and foremost aim is to unwind the DNA molecule without causing any damage to cells or nuclei. James C. Wang was the first to discover a topoisomerase in the 1970s. This class of enzymes opens the DNA strand before replication to avoid and topological problems as supercoiling, knotting, and catenation (Wang, 2002). These enzymes (Topo I and Topo II) generate a nick in the DNA strand, unlock the coiling and then rejoin the strand to make the parent strand integral again (Fig **1.4**). Both these enzymes work in different manners as:

Topo I

Topoisomerase I enzyme was first discovered in *E.coli*, where it generated a nick in the single strand of DNA. After binding an enzyme with a DNA molecule, one strand cuts, concurrently creating a covalent phosphodiester bond with 5′ phosphate on the DNA and a tyrosine residue in the enzyme. The free 3′-OH DNA end is held non-covalently by the enzyme. One DNA strand that has not been cut is passed through the single-stranded break. The cleaved strand is then resealed, making a structure with the help of some chemical bonds as the DNA starting, however, with one less negative supercoil. By this phenomenon, the enzyme eliminates one negative supercoil at a time (Woo, Sun, Cassady, & Snapka, 1999).

Topo II

This enzyme is also called *DNA Gyrase*. It has the capacity to break both strands of a double-stranded DNA molecule, pass another portion of the double-helical structure through the cut, and reseal the cut in a process that consumes ATP. Depending on the DNA substrate, these maneuvers will influence altering a positive supercoil into a negative supercoil or enhancing the number of negative supercoils by 2 (Wang, Caron, & Kim, 1990). The Topo II enzymes (like *E. coli* DNA gyrase) from mammal cells could not increase the super helical density at the expenditure of ATP; likely, no such activity is essential in eukaryotes since binding of histones increases the potential superhelicity. All type II topoisomerases catalyze decatenation and catenation, *i.e.*, the linking and unlinking of two different DNA strands (Pommier, Sun, Shar-yin, & Nitiss, 2016).

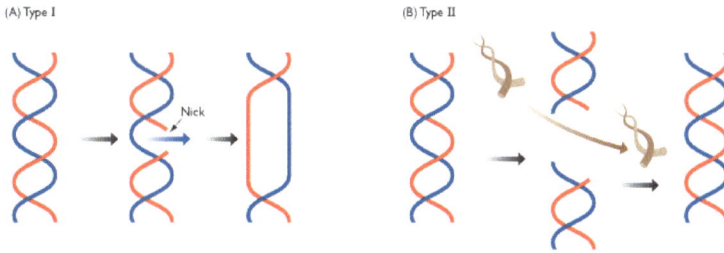

Fig (1.4). Topoisomerases generate cuts in the DNA strand and wide open the topology to start actual duplication.

DNA Polymerases

These are all classes of enzymes that can have all forms of DNA replication. Here one thing is needed to understand that these molecules are not initiators of the process but can extend only the newly built strand at $3'$ end. Kornberg and his colleagues revealed the enzyme DNA polymerase I in 1956. This enzyme is also called Pol I. A Kornberg was awarded Nobel Prize in Medicine or Physiology. Kornberg and Malcolm E. Gefter discovered DNA polymerase II in 1970 (Hubscher, 2010). Various steps of DNA polymerases have been recognized in prokaryotic and eukaryotic cells, which are detailed below;

Priming

DNA replication needs a starting point to initiate the whole process. Primers are short RNA sequences in the nucleus and get attached to the parent strands to initiate replication (Fig. **1.5**). The RNA strands are then altered by DNA polymerase-I for prokaryotes or DNA polymerase δ for eukaryotes. Different types of machinery are used in eukaryotes and prokaryotes for this purpose, and new deoxyribonucleotides are added to fill the gaps where the RNA was existent (Salas, Miller, Leis, & DePamphilis, 1996).

Initiation

Initiation of DNA replication occurs at AT-rich regions of DNA strand called *origin.* These A-T base pairs have a double hydrogen bond which is easier to unzip. After the unwinding of DNA strands with topoisomerases, the Helicase enzyme breaks the hydrogen bonding between the double helix of the strand, converting it into a single strand, which is quite unstable and can be degraded easily. To keep this single strand intact, Single-Strand Binding Proteins get attached with both parent strands at exposed bases to prevent improper ligation. Then RNA primers get attached with both strands providing a site for attachment of *DNA Polymerases* (Dutta & Bell, 1997);(Dhingra & Kaplan, 2016).

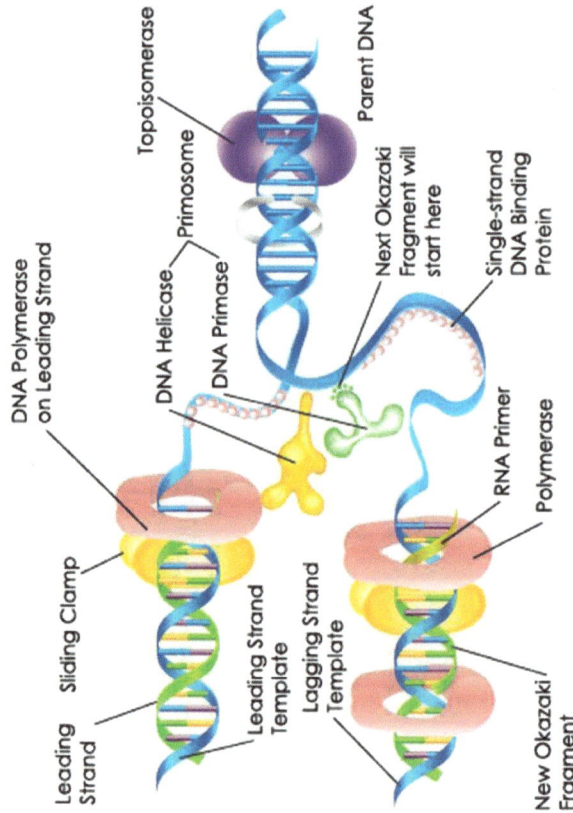

Fig. (1.5). Whole machinery of DNA Replication (www.alyvea.com).

Elongation

As soon as the primers are in place, actual replication starts. DNA polymerase is the enzyme that did all the work of replication. At each primer's 3/OH tail, DNA polymerase binds one dNTPs by removing two phosphates and forming phosphodiester bonds. This continues till the whole bubble is elongated. In the meantime, helicase opens up the helix in front of the growing chain to expose more template strands (Krebs, Lewin, Goldstein, & Kilpatrick, 2013);(Bhagavan & Ha, 2011). The replicated strands keep developing continuously 5' → 3' as helicase makes the template accessible. On the opposite strand, new primers have been added to take advantage of the newly available template.

So, the interaction of opening the helix and synthesizing gDNA 5' → 3' on one strand, although laying down new primers on the other, forwards to the formation of leading and lagging strands.

Leading Strands

The strands are being designed in one bout of uninterrupted DNA synthesis. Leading strands follow the lead of helicase.

Lagging Strands

The strands that begin over and over as new primers are laid down. Synthesis of the lagging strands stops when they reach the 5' end of a primer elsewhere on the strand. The short DNA strands (Okazaki fragments) are formed by lagging DNA synthesis. This is named for the scientist Reiji Okazaki, who discovered these fragments (Hanaoka & Sugasawa, 2016).

Proofreading of Newly Formed Strand

A newly formed DNA strand is proofread to avoid mispairing of nucleotides. Replication is a very rapid process. In eukaryotes speed of replication has been observed at about 2000 bases/minute. This high speed can cause misreading or errors in the newly formed strand; one after every 100,000 bases is incorrect. DNA polymerase constantly checks its work through a process called proofreading. If a wrong base is added, the DNA polymerase will back up and sever the wrong base. The cleavage process is called exonuclease activity, and the calibration process requires DNA polymerase to move 3'→5' instead of the usual 5'→3' direction. DNA proofreading eliminates most of the errors caused by DNA polymerase, resulting in almost no false DNA synthesis. Typically, replication (after proofreading) has a surprisingly low error rate.

Replication of Circular DNA

Mitochondrial or other bacterial circular genomes replicate with a different strategy than mentioned above. There are three key mechanisms involved:

- ➢ Theta
- ➢ Rolling Circle
- ➢ D-Loop

Theta Replication refers to the splitting of the helix and attaining the shape of letter Θ. So after helicase denatures the hydrogen bonds between two strands, replication proceeds bidirectionally and rapidly copying the entire molecule (Fig **1.6**).

Roller loop replication does not require primers because the double-stranded template cleaves at the starting point to provide a free hydroxyl tail to initiate replication. The inner chain is continuously replicated as a leading chain (Fig. **1.6**). At the same time, the broken strands are peeled off. Once sufficient broken chains have been released, the primers are lowered to replicate when the broken strands are stripped from their complement.

D-Loop replication creates a shifted single chain. The helicase opens the double-stranded molecule, lays the RNA primer, and replaces one strand. Then copy the chain around the circle and push the shift as you move. The single intact strand is released and used as a template to synthesize complementary strands (J. A. Lee, Carvalho, & Lupski, 2007).

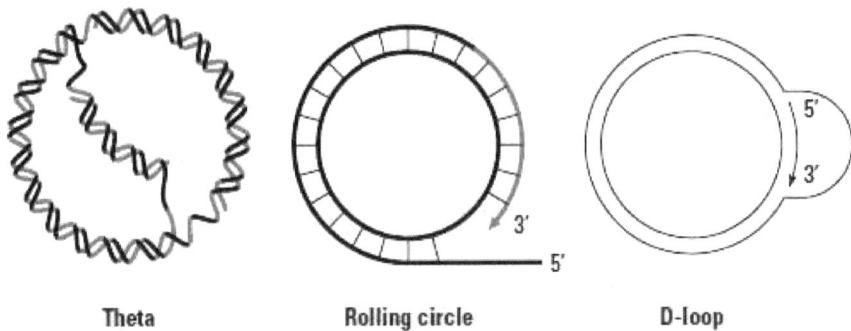

Theta   Rolling circle   D-loop

Fig (1.6). Replication methods of circular DNA

Differentials of Replication in Prokaryotes and Eukaryotes

Prokaryotic replication is quite different than present in Eukaryotes. Carrying complex and advanced cellular components, eukaryotes undergo many advanced forms of replication. If these two organisms are compared for the efficiency of the process, both are equally privileged to conserve the uniqueness and integrity of the genomes of the organisms.

Key differential points in both cell types have been mentioned in Table **2**.

Table 2. Difference of Replication process in prokaryotes and eukaryotes.

Characteristics	Prokaryotes	Eukaryotes
Location	Cytoplasm	Nucleus
Stage	Start of cell division	S-Phase of cell cycle
Origin (ori)	Single	Multiple
Size of ori	100-200 bases	150 bases
Polymerases	**Pol I:** 5' to 3' polymerase, 3' to 5' exonuclease, 5' to 3' exonucleases **Pol III:** 5' to 3' polymerase, 3' to 5' exonuclease	**Pol α:** 5' to 3' polymerase, no exonucleases **Pol δ:** 5' to 3' polymerase, 3' to 5' exonucleases **Pol ε:** 5' to 3' polymerase **Pol γ:** Mitochondrial DNA polymerase (5' to 3')
Telomerase	Absent	Present
Primers	Longer	Shorter
Termination site	Single	Several
Gyrase	Needed	Not needed
Okazaki fragments	1000-2000 bases	100-200 bases

(Table 2) cont.....

Speed	2000 bases /sec	100 bases /sec
Duration	20 min.	18-24 rs.

DNA Damage and Mutations

The replication of DNA might leave some scars, which are overlooked by enzyme machinery involved in DNA duplication. These defects need repair before getting translated into proteins. These changes in DNA are called *Mutations*. DNA mutations refer to any sudden, heritable change in the genotype of an organism rather than through recombination of preexisting genetic variations.

Mutations are either due to misreading during replication process or there may be some other chemical or physical agents causing these variations in the sequence of DNA bases (Friedberg, Walker, Siede, & Wood, 2005). These causative agents are called *Mutagens*. Some of the mutagenic agents are given below:

Chemical Mutagens

These are classified into three major groups on the basis of their specific reaction with DNA molecules.

Base Analog

It is a molecule whose structure is close enough to the natural base that is sometimes incorporated into DNA in place of the natural base.

Alkylating Agents

They react with DNA by adding ethyl or methyl groups to the bases.

Acridine Dyes

They are a class of chemicals that intercalate between the bases of DNA.

Physical Mutagens

Ionizing Radiations

Natural radiations from radioactive substances induce spontaneous mutations. These include X-rays, gamma rays, neutrons, beta and alpha particles, ultraviolet radiations, *etc*.

Environmental Mutagens

This class includes air and water pollutants, food additives, preservatives, and agricultural chemicals.

Types of Mutations

All different types of mutations are due to three main errors in the DNA complementary strand.

- Deletion
- Insertion
- Substitution

Due to these errors following types of variations can occur in the DNA:

Point Mutation

When the only base is substituted at any position to some other base.

Synonymous Mutation

When a change in nucleotide is not transferred to protein. It is also called a silent mutation. This mutation is due to redundancy in genetic code.

Non-Synonymous Mutation

When a mutation in DNA strand cause a change of amino acid in translated protein.

Frame-shift Mutation

If deletion or insertion of one base causes a change in the reading frame of the protein.

Somatic Mutation

When mutation occurs in the somatic cell, which can cannot be transferred to the next generation.

Germ-line Mutations

When mutations occur in the germ cells and are transferred to the next generation.

Missense Mutations

When mutation causes a change of amino acid in the protein polypeptide.

Non-sense Mutations

When a mutation causes any early STOP codon and resulted in truncated protein.

Spontaneous Mutations

Which arise randomly in nature without any readily apparent cause.

Induced Mutations

Which occur in response to an obvious externally applied agent.

DNA Repair

Although DNA is a highly stable molecule, spontaneous changes occur even under normal cell conditions, and mutations can result if they are not repaired.

Being a double helix molecule, DNA is well suited for repair as it carries template copy to the safe storage of genetic information, which can restore the corrected nucleotide sequence to the damaged strand. Each cell contains multiple DNA repair systems, each with its own enzymes and a preference for the type of damage identified (Fig. **1.7**).

DNA Damage in Single Strand

Single-stranded DNA damages can be repaired by three major pathways:

a)   Base excision repair
b)   Nucleotide excision repair
c)   Mismatch repair

a)   Base Excision Repair

This process involves the removal of the damaged or mismatched base by a specialized enzyme called *DNA glycosylase*. This enzyme is encoded by eight different genes that can identify and remove a specific type of base damage.

This removal of a base produces a gap in the strand filled by *DNA polymerase beta*. This enzyme substitutes the incorrect base with the corrected one. Then strand is joined back by Ligases.

b)   Nucleotide Excision Repair

Also called NER is different from BER in many ways. It uses different enzymes to splice out the DNA, and it may also cause the removal of more than one base from the DNA strand.

The main steps are: firstly, the damage is identified by one or more protein factors that came together at the location. Then DNA is wide opened to create a bubble, and transcription Factor IIH, TFIIH cut the defective base from the newly formed strand. The intact strand acts as a template, and a complementary base is added to the defected strand by *DNA polymerase delta* and *epsilon*. Then *DNA ligase* repairs the backbone by a phosphodiester bond.

c)   *Mismatch Repair*

The mismatch repairs are chain-specific. The newly synthesized (sub)chain usually includes errors during DNA synthesis. To begin the repair, the mismatch repair machine separates the newly synthesized chain from the template (parent). The mismatch repair enzyme detects any difference between the template and the newly synthesized strand, so they cut off the wrong base and use the template strand as a guide to insert the correct base (Chen, 2013).

DNA Damage in Doubles Strand

When the two strands of the double helix break, a potentially dangerous type of DNA damage occurs, and no complete template strand can serve as a template for repair. This type of break is caused by ionizing radiation, oxidants, replication errors, and certain metabolites in the cell. If these lesions are not repaired, they quickly cause the chromosomes to break down into smaller pieces. However, two different mechanisms have been developed to improve potential damage.

a) Non-homologous end-joining
b) Homologous end-joining

a) *Non-homologous End-Joining*

This type of DNA repair causes the loss of some parts of the genome as no complementary strand is available to be used as a template. In this method, the broken ends of DNA are joined together by ligases. This end-joining mechanism is also termed as an immediate solution to the massive broken regions in the double strand of DNA. As an outcome of this type of repair, the sequence of the DNA is changed (a mutation), with so few mammalian genomes encoding proteins; this mechanism is clearly an acceptable solution to the problem of maintaining chromosome integrity.

b) *Homologous End-Joining*

This type of repair is more effective for lesions in double-stranded DNA. A diploid cell carries two copies of a chromosome or double-helix DNA molecule. So if one copy is damaged, then through recombination, transfer of intact DNA molecule occurs at the place of damage. This process requires several proteins that can recognize the areas of DNA matching between two chromosomes and bring them together. Then replication machinery uses an intact strand as a template to add complementary bases at the damaged site, repairing it with no changes in the nucleotides sequence (Hanaoka & Sugasawa, 2016).

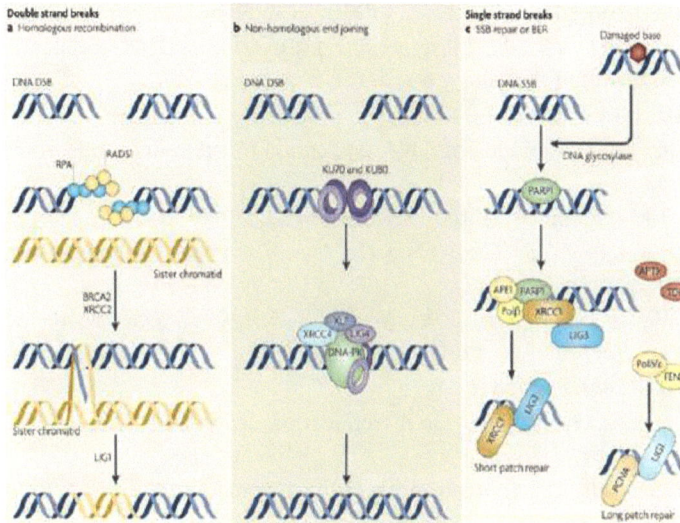

Fig. (1.7). DNA repair mechanisms in the cell. (www.nature.com).

CONCLUSIONS

This chapter describes major events in the historical timeline of genomics. A complete structure of DNA is described in detail. Molecular organization of the genome, the evolution of a single cell into a complex form of life is also described. Mechanism of replication of genome, DNA damage, mutations, and repair mechanism are discussed.

REFERENCES

Barbieri, M. (2007). *The Codes of Life: The Rules of Macroevolution.* Springer Science & Business Media.

Bernardi, G. (1995). The human genome: organization and evolutionary history. *Annu. Rev. Genet., 29*(1), 445-476.

http://dx.doi.org/10.1146/annurev.ge.29.120195.002305 PMID: 8825483

Bhagavan, N.V., Ha, C-E. (2011). *Essentials of Medical Biochemistry: with Clinical Cases.* Academic Press.

Chen, C. (2013). *New research directions in DNA repair.* BoD–Books on Demand.
http://dx.doi.org/10.5772/46014

Dahm, R. (2005). Friedrich Miescher and the discovery of DNA. *Dev. Biol., 278*(2), 274-288.
http://dx.doi.org/10.1016/j.ydbio.2004.11.028 PMID: 15680349

Dhingra, N., Kaplan, D.L. (2016). Introduction to Eukaryotic DNA Replication Initiation.*The Initiation of DNA Replication in Eukaryotes.* Springer.
http://dx.doi.org/10.1007/978-3-319-24696-3_1

Dutta, A., Bell, S.P. (1997). Initiation of DNA replication in eukaryotic cells. *Annu. Rev. Cell Dev. Biol., 13*(1), 293-332.
http://dx.doi.org/10.1146/annurev.cellbio.13.1.293 PMID: 9442876

Fridell, R. (2005). *Decoding Life: Unraveling The Mysteries of the Genome.* Twenty-First Century Books.

Friedberg, E.C., Walker, G.C., Siede, W., Wood, R.D. (2005). *Dna Repair and Mutagenesis.* American Society for Microbiology Press.
http://dx.doi.org/10.1128/9781555816704

Hanaoka, F., Sugasawa, K. (2016). DNA replication, recombination, and repair: Molecular mechanisms and pathology.

Hand, R. (1978). Eucaryotic DNA: organization of the genome for replication. *Cell, 15*(2), 317-325.
http://dx.doi.org/10.1016/0092-8674(78)90001-6 PMID: 719745

Hartl, D.L. (2014). *Essential genetics: A genomics perspective.* Jones & Bartlett Publishers.

Hubscher, U. (2010). *DNA Polymerases: Discovery, Characterization, and Functions in Cellular DNA Transactions.* World Scientific.
http://dx.doi.org/10.1142/7667

Krebs, J.E., Lewin, B., Goldstein, E.S., Kilpatrick, S.T. (2013). *Lewin's essential genes.* Jones & Bartlett Publishers.

Lee, J.A., Carvalho, C.M., Lupski, J.R. (2007). A DNA replication mechanism for generating nonrecurrent rearrangements associated with genomic disorders. *Cell, 131*(7), 1235-1247.
http://dx.doi.org/10.1016/j.cell.2007.11.037 PMID: 18160035

Lee, T.F. (2013). *The Human Genome Project: Cracking the genetic code of life.* Springer.

Maddox, B. (2002). *Rosalind Franklin: The dark lady of DNA.* HarperCollins New York.

Margulis, L., Chapman, M.J. (1998). Endosymbioses: cyclical and permanent in evolution. *Trends Microbiol., 6*(9), 342-345.
http://dx.doi.org/10.1016/S0966-842X(98)01325-0 PMID: 9778725

Margulis, L., Fester, R. (1991). *Symbiosis as a source of evolutionary innovation: speciation and morphogenesis.* Mit Press.

Müller, C.A., Hawkins, M., Retkute, R., Malla, S., Wilson, R., Blythe, M.J., Nakato, R., Komata, M., Shirahige, K., de Moura, A.P., Nieduszynski, C.A. (2014). The dynamics of genome replication using deep sequencing. *Nucleic Acids Res., 42*(1), e3-e3.
http://dx.doi.org/10.1093/nar/gkt878 PMID: 24089142

O'Malley, M.A. (2015). Endosymbiosis and its implications for evolutionary theory. *Proc. Natl. Acad. Sci. USA, 112*(33), 10270-10277.

http://dx.doi.org/10.1073/pnas.1421389112 PMID: 25883268

O'Reilly, R.K., Turberfield, A.J., Wilks, T.R. (2017). The evolution of DNA-templated synthesis as a tool for materials discovery. *Acc. Chem. Res., 50*(10), 2496-2509.

http://dx.doi.org/10.1021/acs.accounts.7b00280 PMID: 28915003

Pommier, Y., Sun, Y., Huang, S.N., Nitiss, J.L. (2016). Roles of eukaryotic topoisomerases in transcription, replication and genomic stability. *Nat. Rev. Mol. Cell Biol., 17*(11), 703-721.

http://dx.doi.org/10.1038/nrm.2016.111 PMID: 27649880

Ramani, V., Shendure, J., Duan, Z. (2016). Understanding spatial genome organization: methods and insights. *Genomics Proteomics Bioinformatics, 14*(1), 7-20.

http://dx.doi.org/10.1016/j.gpb.2016.01.002 PMID: 26876719

Rippe, K. (2012). *Genome Organization and Function in the Cell Nucleus.* John Wiley & Sons.

Salas, M., Miller, J. T., Leis, J., DePamphilis, M. L. (1996). Mechanisms for priming DNA synthesis. *DNA Replication in Eukaryotic Cells,* 131-176.

Sinden, R.R. (1994). *DNA structure and function.* Gulf Professional Publishing.

Wachsmuth, M., Caudron-Herger, M., Rippe, K. (2008). Genome organization: balancing stability and plasticity. *Biochimica Et Biophysica Acta (Bba)-. Molecular Cell Research, 1783*(11), 2061-2079.

Waga, S., Stillman, B. (1998). The DNA replication fork in eukaryotic cells. *Annu. Rev. Biochem., 67*(1), 721-751.

http://dx.doi.org/10.1146/annurev.biochem.67.1.721 PMID: 9759502

Wang, J.C. (2002). Cellular roles of DNA topoisomerases: a molecular perspective. *Nat. Rev. Mol. Cell Biol., 3*(6), 430-440.

http://dx.doi.org/10.1038/nrm831 PMID: 12042765

Wang, J.C., Caron, P.R., Kim, R.A. (1990). The role of DNA topoisomerases in recombination and genome stability: a double-edged sword? *Cell, 62*(3), 403-406.

http://dx.doi.org/10.1016/0092-8674(90)90002-V PMID: 2165864

Woo, S.H., Sun, N-J., Cassady, J.M., Snapka, R.M. (1999). Topoisomerase II inhibition by aporphine alkaloids. *Biochem. Pharmacol., 57*(10), 1141-1145.

http://dx.doi.org/10.1016/S0006-2952(99)00018-0 PMID: 11230801

CHAPTER 2

The Spectrum of Genomics

Abstract: This chapter proposes a brief description of: From Gene to Protein; Regulation of Gene Expression; Value of Diversity; Gene evolution: Divergence of sequences and structures within and between species; Economic and national needs of biodiversity; Conservational efforts for sustainable ecosystem.

Keywords: Biodiversity, Central dogma, Gene evolution, Genetic conservation.

The world of genomics deals with life philosophy: from one-dimensional DNA molecules to three-dimensional protein polypeptides. This concept defines the whole central dogma of molecular biology: how DNA can be converted into a functional product, which is the protein in most instances. To understand this dynamic shift among functional entities, there is a need to illustrate the step-by-step process of RNA formation and, ultimately, protein synthesis.

FROM GENE TO PROTEIN

The central rule of molecular biology is the interpretation of the flow of genetic information within biological systems. It is often referred to as "DNA makes RNA and RNA produce proteins," although this is not its original meaning. Frances Creek first proposed this in 1958;

"Central dogma: This means that once the 'information' enters the protein, it cannot reappear. In more detail, the transfer of information from nucleic acids to nucleic acids or from nucleic acids to proteins is possible, but from proteins to proteins or from proteins to nucleic acids transfer is not possible. Information here means precisely determining the sequence, the base in a nucleic acid or an amino acid residue in a protein".

The concept of central dogma was re-stated in a Nature paper published in 1970;

"The core rule of molecular biology is to handle detailed residual shifts in sequence information. It states that this information cannot be transferred from proteins to proteins or nucleic acids."

Maryam Javed, Asif Nadeem & Faiz-ul Hassan

The second version of the central doctrine is very popular but not correct. This is the simple DNA→RNA→protein pathway published by James Watson in the first edition of The Gene Biology of the Gene (1965). Watson's version is different from Crick's version because Watson describes the two-step (DNA→RNA and RNA→protein) pathway as a central rule (Fig. **1**). Although Crick's original dogma still works today, Watson's version does not.

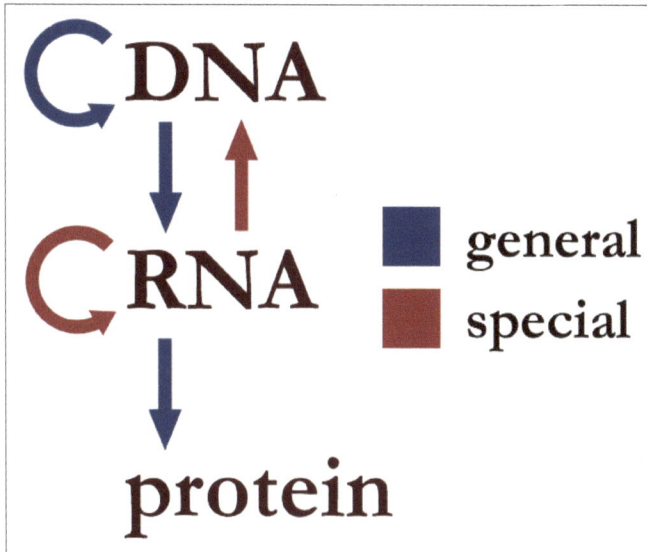

Fig. (1). Information Flow in Biological Systems.

Philosophy of Life: From One Dimension to Three Dimensions

The discovery of the phenomenon of protein synthesis required the development of cell-free extracts capable of carrying on vital synthetic steps. These were first effectively developed in 1953 by Paul C. Zamecnik and his colleagues. The key to this was tagging amino acids with radioactive dyes and high-quality preparative ultracentrifuges for cellular fractionation.

A few years later, Zamecnik and Mahlon B. Hoagland embarked on a significant new discovery. Before they were incorporated into proteins, the amino acids were first combined with what we now call transfer RNA (tRNA), which accounts for nearly 10% of the total number of cell's RNA count (Farabaugh, 2012).

Given the presence of 20 amino acids but only 4 bases, a few nucleotide groups must specify a given amino acid in some way. However, the two groups specified only 16 (4 x 4) amino acids. So, starting in 1954, scientists started thinking seriously about what the genetic code might look like. Triplets were found to be the most logical answer (4 x 4 x 4). But which particular three base set determines which specific amino acid can be learned by biochemical analysis (Griffiths & Stotz, 2013). The genetic code was completed in 1966, showing that 61 of the 64 possible substitution groups correspond to amino acids, and most amino acids are encoded by more than one nucleotide triplet (Fig. **2**).

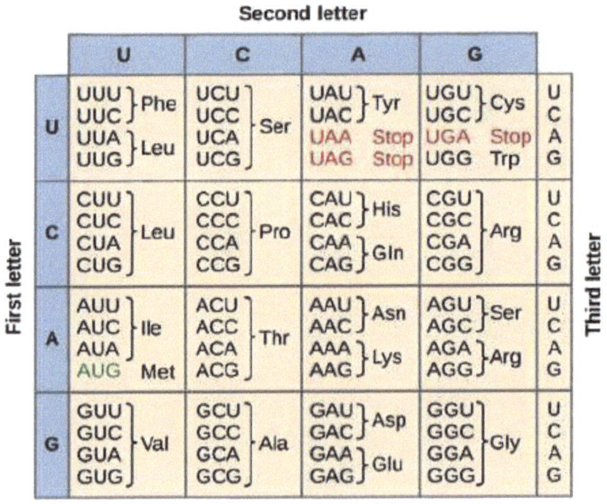

Second letter

	U	C	A	G	
U	UUU UUC } Phe UUA UUG } Leu	UCU UCC UCA UCG } Ser	UAU UAC } Tyr UAA Stop UAG Stop	UGU UGC } Cys UGA Stop UGG Trp	U C A G
C	CUU CUC CUA CUG } Leu	CCU CCC CCA CCG } Pro	CAU CAC } His CAA CAG } Gln	CGU CGC CGA CGG } Arg	U C A G
A	AUU AUC } Ile AUA AUG Met	ACU ACC ACA ACG } Thr	AAU AAC } Asn AAA AAG } Lys	AGU AGC } Ser AGA AGG } Arg	U C A G
G	GUU GUC GUA GUG } Val	GCU GCC GCA GCG } Ala	GAU GAC } Asp GAA GAG } Glu	GGU GGC GGA GGG } Gly	U C A G

First letter (left vertical axis) — Third letter (right vertical axis)

Fig. (2). The Genetic Code.

mRNA Synthesis in Transcription

Only 4% of the total cellular RNA is mRNA. This shows the large variation in length, depending upon the polypeptides for which they code. To transcribe a gene, RNA polymerases proceed through a series of well-defined steps which are grouped into three phases: *initiation, elongation* and *termination* (Fig. **3**).

Initiation: A promoter is a conserved sequence of DNA, which allows the binding of RNA polymerase and many transcriptions factors to initiate the transcription process. After the promoter-polymerase complex formation, there are a series of structural changes in the DNA, including unwinding and formation of transcription bubble. As in replication initiation, here as well, the DNA bases are denatured, and the double-strand is converted into a single strand. The direction of polymerization is $5'$ to $3'$ as it also occurs in replication. That is, a new ribonucleotide is added to the 3/end of the growing strand. Unlike replication, only one DNA strand acts as a template for constructing an RNA strand, which further serves as a template for protein formation (Kais, 2018).

Elongation: Once RNA polymerase synthesizes a small stretch of RNA (approximately 10 bases), it transitions to an extended phase. This conversion requires a further conformational change of the polymerase to hold the template more firmly. During the extension and the catalysis of RNA synthesis, the enzyme performs an impressive task. It unfolds the front DNA and reanneals it, separating the growing RNA strand from the template as the template moves and performing proofreading (Green & Noller, 1997).

Termination: Once the polymerase transcribes the length of the gene, it must stop and release the RNA product. This step is called termination. In some cells, there is a specific, well-characterized sequence that triggers termination; in other respects, it is not clear what indicates that the enzyme stops transcription and dissociates from the template.

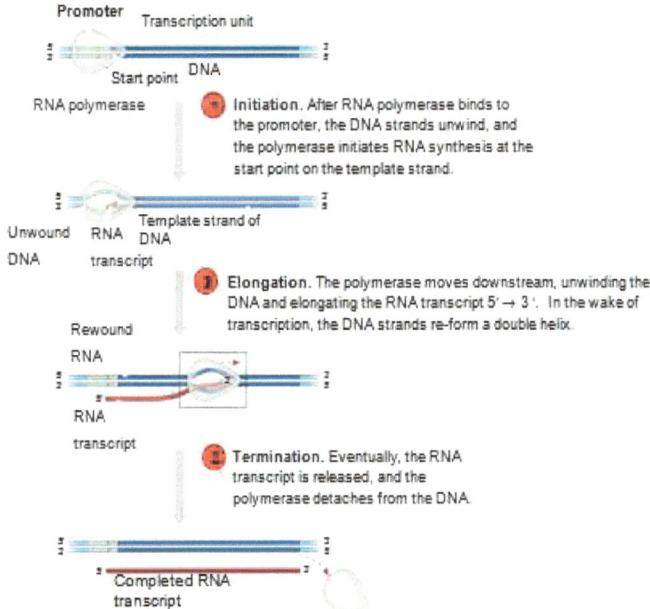

Fig. (3). Transcription Events.

Post Transcriptional Modifications

Once the initial transcript is formed in eukaryotes, RNA proceeds for various processing steps, increasing its stability and functional efficiency. After these changes, the final RNA molecule from the nucleus travels to the cytoplasm, where it is translated into protein polypeptide. These treatment events include 5/terminal capping of RNA; splicing and polyadenylation are also known as tailing of the 3/end of RNA. The most complicated of these is splicing, which removes non-coding introns from RNA to produce mature mRNA (Kais, 2018).

5/ Capping: The first step in RNA modification is capping or methylation. This involves the addition of altered guanine bases at the 5/end of the RNA. Specifically, the base is methylated. When the RNA is still only about 20-40 nucleotides long, it is capped. Upon capping, dephosphorylation of Ser5 within the tail repeat results in dissociation of the capping mechanism, and further phosphorylation results in the recruitment of machinery required for RNA splicing.

3/ Tailing: The final RNA processing event, polyadenylation of the 3/end of mRNA, is closely related to transcription termination. Once the polymerase reaches the end of the gene, it encounters termination sequences which, upon transcription, trigger the transfer of polyadenylation to the RNA, resulting in three events: the cleavage of the transcript at its 5' end, addition of many adenine residues at 3' and the termination of transcription by the polymerase (Wu, 2013);(Puri, 2018).

Translational Machinery

The machine responsible for translating the messenger RNA language into proteins consists of four major components: mRNAs, tRNAs, and ribosomes. Together, these components complete the extraordinary task of translating the code of four basic alphabets (A, T, C, G) into 20 amino acids.

mRNA: Messenger RNA contains at least one open reading frame. The number of ORFs per mRNA is different between eukaryotes and prokaryotes. Eukaryotic mRNAs always contain a single ORF. In contrast, prokaryotic mRNAs typically include two or more ORFs and thus can encode multiple polypeptide chains. An mRNA having multiple ORFs is called a polycistronic, and having a single ORF is referred to as a single cistron.

tRNAs: There are many types of tRNA molecules, but each is attached to a specific amino acid and a particular recoginized codon. tRNAs are between 75-95 nucleotide lengths.

Ribosomes: These are macromolecules that direct the synthesis of proteins. Ribosomes are composed of two RNAs and proteins called size subunits. They provide primary sits for attachment of mRNA and further protein synthesis (Rodnina, Wintermeyer, & Green, 2011).

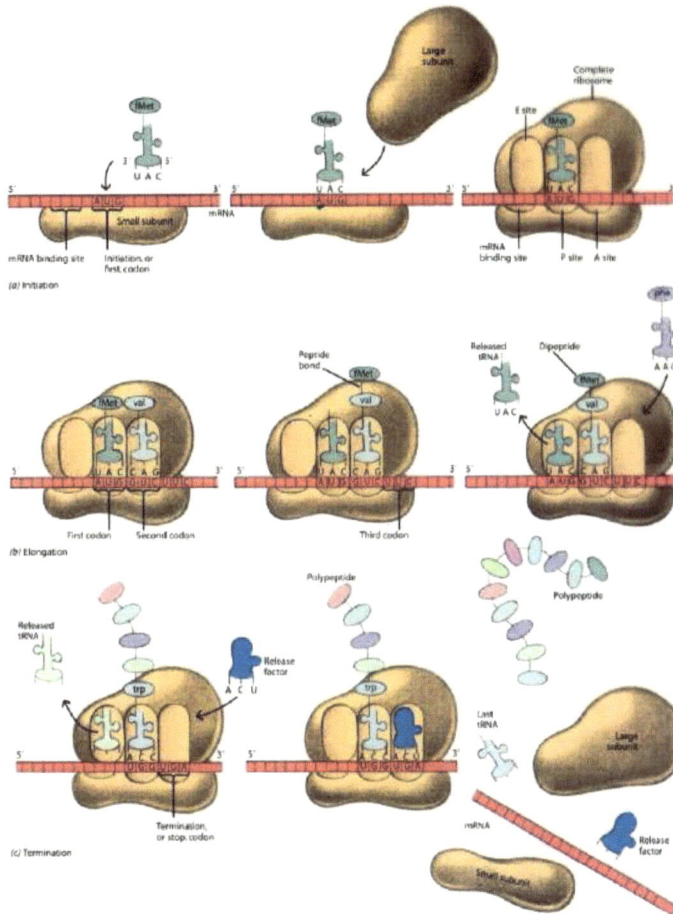

Fig. (4). Eukaryotic Protein Translation.

Translation

As a translation initiation, the ribosome assembles with a charged initiator tRNA in the P site to initiate polypeptide synthesis (Fig.**3**). In order to properly add each amino acid, three key events occurred. First, the correct aminoacyl-tRNA is loaded into the A site of the ribosome, as indicated by the A-site codon. Next, a peptide bond is formed between the aminoacyl-tRNA at the A site and the peptide chain linked to the peptidyl-tRNA at the p site. Third, the peptidyl-tRNA and its associated codons generated in the A site must be translocated to the P site so that the ribosome can undergo another cycle. These three steps are termed as;

Initiation: It is the start of translation.

Elongation: It is the growing of the peptide chain.

Termination: Ending of translation and detachment of all subunits.

The translation error rate is between 10-3 and 10-4. This is just that I am not correct for every 1000 amino acids incorporated into the protein. The ultimate basis for selecting the correct aminoacyl-tRNA is base pairing between the charged tRNA and the codon shown in the ribosome A site.

Post Translation Modifications

Post protein synthesis modifications are necessary to make it a fully functional product. The main reasons for this are enhancing the stability of protein, biochemical activity (activity regulation), protein targeting (protein localization), and protein signaling (Protein-Protein interaction, cascade amplification) (Puri, 2018); (Trachsel, 1991). Some of the major modifications are given in Table **1** below;

Table 1. Types of PTM in Eukaryotes.

Modification	Description
ACETYLATION	N-terminal of some residues and side chain of lysine
AMIDATION	Generally at the C-terminal of a mature active peptide after oxidative cleavage of last glycine
BLOCKED	Unidentified N- or C-terminal blocking group
FORMYLATION	Generally of the N-terminal methionine
GAMMA-CARBOXYGLUTAMIC ACID	Of glutamate
HYDROXYLATION	Generally of asparagine, aspartate, proline or lysine
METHYLATION	Generally of N-terminal phenylalanine, side chain of lysine, arginine, histidine, asparagine or glutamate, and C-terminal cysteine
PHOSPHORYLATION	Of serine, threonine, tyrosine, aspartate, histidine or cysteine, and, more rarely, of arginine
PYRROLIDONE CARBOXYLIC ACID	N-terminal glutamine which has formed an internal cyclic lactam. This is also called 'pyro-Glu'. Very rarely, pyro-Glu can be produced by modification of a N-terminal glutamate
SULFATION	Of tyrosine, serine or threonine

Regulation of Gene Expression

Studies on gene action regulation are centered primarily on *E.coli* and its enzymes. The step following the initiation of transcription can regulate gene expression. For example, the modulation can be at the level of transcriptional elongation. Three cases can be considered here: attenuation of the trp gene and resistance termination of the N and Q proteins of phage lambda. The trp gene encodes the enzyme required to synthesize the amino acid tryptophan. One way in which amino acids control the expression of these genes is attenuation. If tryptophan is present in the cell (in the form of trp tRNA), the transcript initiated at the trp promoter is preceded by the transcription of the structural gene. The lambda proteins N and Q are loaded onto RNA polymerase and transcription is initiated at specific promoters in the phage genome. Once modified in this way, the enzyme can pass through specific transcriptional terminator sites that would otherwise block the expression of downstream genes.

In addition to transcription, gene regulation proceeds at the translational level. The correct expression of the ribosomal protein gene presents an interesting regulatory problem for the cell. Each ribosome contains about 50 different proteins and must be prepared at the same rate. In addition, the rate at which a cell produces protein, as well as the amount of ribosome it requires, is closely related to the rate of growth of the cell; changes in growth conditions quickly result in an increase or decrease in the rate of synthesis of all ribosomal components (Harding & Crabbe, 1991);(Perdew, Heuvel, & Peters, 2008).

Value of Diversity

Homologous recombination occurs in all organisms, allowing for the exchange and recombination of genes along the chromosome and repairing the broken DNA strand and the folded replication fork. These events eventually created new forms of genes and helped to rearrange existing chromosome structures. Recombination involves the cleavage and reconnection of DNA molecules. The double-stranded repair pathway of homologous recombination well describes many recombination events. With this model, the exchange initiation requires that one of the two homologous DNA molecules have a double-strand break. A DNA degrading enzyme carries out the end of DNA fragmentation to produce a single-stranded DNA fragment. These single-stranded regions are involved in DNA pairing with homologous partner DNA. Once pairing occurs, the two DNA molecules are joined by a branched structure in the DNA called a Holliday junction. Cutting DNA at the Holliday junction resolves the junction and terminates the reorganization. Holliday junctions can be cut in two ways. One method produces a cross-product in which regions from two parental DNA molecules are now covalently linked. An alternative method of cleavage ligation produces recombinant DNA fragments but does not result in a crossover.

Cells encode enzymes that catalyze all the steps in homologous recombination. Key enzymes are the strand exchange proteins. During meiosis, recombination is essential for the proper homologous pairing of chromosomes before the first nuclear division. Therefore, recombination is highly regulated to ensure it occurs on all chromosomes.

Gene Evolution: Divergence of Sequences and Structures within and between Species

Gene Duplication

The increase in the number of genes in vertebrates is mainly due to gene duplication. But how to increase the copy number of certain genes lead to an increase in morphological diversity? There are two ways to do this. First, the traditional view is that ancestral genes produce multiple genes by replication, and the coding regions of new genes undergo mutations. This replication process typically does not produce new genes that encode proteins with novel functions. Instead, it produces genes encoding related proteins with slightly different activities. The second approach to produce diversity is by the variation in repeat regions. According to this model, the repeated gene does not necessarily have a new function, but a new regulatory DNA sequence is obtained. This allows different copies of the gene to be expressed in different patterns in the developing organism (Ohno, 2013);(Pontarotti, 2016).

The high degree of conservation of genes recently discovered in different animals has led to concerns about changes in gene expression as a general mechanism for generating evolutionary diversity. The importance of this mechanism highlights the morphological changes caused by gene-expressing genes in new places during Drosophila development.

There are two ways of duplication in the genome: replicative transposition (gene duplication *via* DNA intermediate molecule) and retrotransposition (gene duplication *via* RNA intermediate molecule) (Dittmar & Liberles, 2011);(Beurton, Falk, & Rheinberger, 2000).

Transposition via DNA Intermediate

The first step in replicating a transposition is to assemble transposase proteins at both ends of the transposon DNA to create a transposon. The next DNA is cleaved at the end of the transposon DNA. This reaction is catalyzed by a transposase in the transposome. The enzyme introduces a nick into the DNA at each of the two junctions between the transposon sequence and the flanking host DNA. The 3/OH end of the transposon DNA is then ligated to the target DNA site by a DNA strand transfer reaction. The intermediate produced by the DNA strand itself is in a bi-branched DNA molecule. The two DNA branches within the intermediate have the structure of a replication fork. After DNA strand transfer, DNA replication proteins from host cells can be assembled at these forks and continue to replicate. This reaction produces two copies of the transposon DNA.

Retrotransposition

This cycle of transposition starts with the transcription of DNA sequence into RNA. The RNA is then reverse transcribed to produce a double-stranded DNA molecule. This DNA molecule is called cDNA and does not contain any flanking host DNA sequences. It is a cDNA recognized by the integrase protein, which helps it reinsert into random locations in the DNA and cause duplication of existing genes. The phenomenon of transformation can lead to the formation of new genes and the dysfunction of existing genes by making Pseudo (Krebs, Lewin, Goldstein, & Kilpatrick, 2013).

Orthologous and Paralogous Genes

Similar genes may have multiple forms in different species due to Molecular to the common ancestral sources. Even multiple genes in one species are the outcome of the gene duplication events. This causes many genes to share some of the physical and functional consequences (Lupski & Stankiewicz, 2007). Orthologous genes are descendants from a common ancestress, while paralogous genes diverge from one another within a species (Fig. **5**).

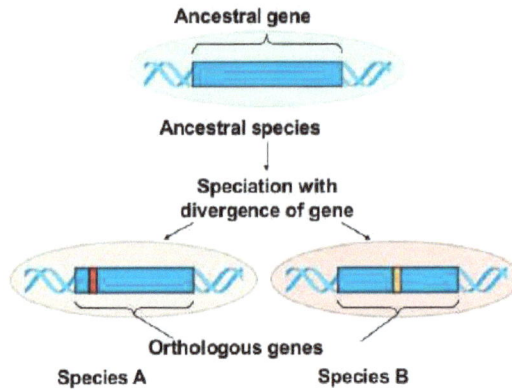

Fig. (5). Orthologous and Paralogous Genes.

Economic and National Needs of Biodiversity

The molecular need for diversity and variation is direly controlling species adaptation to the ever-changing environment of the earth. The genetic differences between races are the reflection of the environmental differences that exist between territories. Mutations arise as a random event all the time. Due to environmental demands, some mutations are favored in one locality but selected against in another. From such differences in the selection, a population is established with specified genetic characteristics that are passed on next generations (Pearce & Moran, 2013).

Conservational Efforts for Sustainable Ecosystem

Protecting, restoring, and sustainably using biodiversity can provide a viable solution to a range of social challenges. The main methods include genetic-based protection, species-based protection, ecosystem-based protection, and landscape-level protection methods. It can be seen from this synthesis that the ecosystem approach is considered to be the best way to protect biodiversity.

To provide the natively needed climate and environment to our flora and fauna is our major responsibility. Due to extensive urbanization, many of our species have lost their home tracts and are in grieving danger of extinction. In the last few years, we have witnessed the extinction of white rhinoceros and many other plant and animal species. There is a need of collective and collaborative efforts of ecologists, geneticists and wildlife conservationists to restore the species conservation to maintain the biodiversity of our planet (Bennett, 2004).

CONCLUSION

This chapter presents a detailed description of information flow in a biological system from gene to protein. Moreover, details regarding the philosophy of life, regulation of gene expression, value of diversity, gene evolution, and economic and national needs of biodiversity are presented. The need for conservational efforts for a sustainable ecosystem is also narrated.

REFERENCES

Bennett, G. (2004). *Integrating Biodiversity Conservation and Sustainable Use: Lessons Learned From Ecological Networks.* IUCN.

Dittmar, K., Liberles, D. (2011). *Evolution After Gene Duplication.* John Wiley & Sons.

Farabaugh, P.J. (2012). *Programmed Alternative Reading of the Genetic Code: Molecular Biology Intelligence Unit.* Springer Science & Business Media.

Green, R., Noller, H.F. (1997). Ribosomes and translation. *Annu. Rev. Biochem., 66*(1), 679-716. http://dx.doi.org/10.1146/annurev.biochem.66.1.679 PMID: 9242921

Griffiths, P., Stotz, K. (2013). *Genetics and Philosophy: An Introduction.* Cambridge University Press. http://dx.doi.org/10.1017/CBO9780511744082

Harding, J.J., Crabbe, M.J.C. (1991). *Post-Translational Modifications of Proteins.* CRC Press.

Kais, G. (2018). *Transcriptional and Post-transcriptional Regulation.* BoD–Books on Demand.

Krebs, J.E., Lewin, B., Goldstein, E.S., Kilpatrick, S.T. (2013). *Lewin's Essential Genes.* Jones & Bartlett Publishers.

Lupski, J.R., Stankiewicz, P.T. (2007). *Genomic Disorders: The Genomic Basis of Disease.* Springer Science & Business Media.

Ohno, S. (2013). *Evolution By Gene Duplication.* Springer Science & Business Media.

Pearce, D., Moran, D. (2013). *The Economic Value of Biodiversity.* Routledge. http://dx.doi.org/10.4324/9781315070476

Perdew, G.H., Heuvel, J.P.V., Peters, J.M. (2008). *Regulation of Gene Expression.* Springer.

Pontarotti, P. (2016). *Evolutionary biology: Convergent Evolution, Evolution of Complex Traits, Concepts and Methods.* Springer. http://dx.doi.org/10.1007/978-3-319-41324-2

Puri, D. (2018). *Textbook of Medical Biochemistry E-BK.* Elsevier Health Sciences.

Richards, R. J. (2000). *The Concept of the Gene in Development and Evolution: Historical and Epistemological Perspectives.* Cambridge University Press.

Rodnina, M.V., Wintermeyer, W., Green, R. (2011). *Ribosomes Structure, Function, and Dynamics.* Springer Science & Business Media.

http://dx.doi.org/10.1007/978-3-7091-0215-2

Trachsel, H. (1991). *Translation in eukaryotes.* Crc Press.

Wu, J. (2013). *Post-transcriptional gene regulation: RNA processing in Eukaryotes.* John Wiley & Sons.

http://dx.doi.org/10.1002/9783527665433

CHAPTER 3

State-of-the-Art Techniques in Genomic Era

Abstract: This chapter proposes a brief description of DNA Polymerization; Genomic Finger Printing; DNA Sequencing; Comparative, Structural and Functional Genomics: A Triode; Proteomics; Metabolomics; Microarrays & Transcriptomics; Genome-wide association study.

Keywords: Metabolomics, Phylogeny, Polymerization, Proteomics, Sequencing, transcriptomics.

The modern world of molecular biology relies on various State-of-the-Art technologies to draw fruitful conclusions from biological data that help understand the much complex phenomenon at the molecular and cellular level. The combination of dry and wet laboratory techniques needs expertise and skills attained by a better understanding of its principles. Some of the latest imperative lab techniques have been featured in this chapter.

DNA POLYMERIZATION

As described in the previous chapter, DNA is a double helical molecule structured in a supercoiled manner to accomplish all of the much-needed activities of the cell. In normal circumstances, DNA polymerizes in the cell to pass the genetic information to the next generations. After the discovery of DNA, the major breakthrough that revolutionized every aspect of molecular biology was the *in-vitro* optimization of DNA polymerization. More than 20,000 researchers cited publication by Kary Mulis, and even nowadays is the single most imperative laboratory technique in cell biology, biotechnology, and molecular biology.

Polymerization in DNA outside the body requires many chemical players and temperature variations to open up the double helix, find a target hot spot for polymerization and finally extend at $3^{/}$ end to the required length. The efficiency of this process is comparable with natural replication that happens inside the cell. But the actual technique is a bit manipulated when conducted in the laboratory. Details of this major innovation in molecular biology have been given under:

Maryam Javed, Asif Nadeem & Faiz-ul Hassan

Principle of Polymerization

As the name indicates, PCR is a chain reaction that can create multiple copies of targetted DNA exponentially. This is not as simple as the given one-sentence statement. One of the many challenges is to identify the target region by using single-stranded short DNA sequences as priming sites. These are also called initiators of polymerization. After binding these primers, the extension of the parent strand is done bi-directionally, and daughter copies are generated. This cycle is repeated at the desired number of times to get multiple replicas of the same region, being further used to study mutations, species evolution, or identify functionally significant genomic regions (Chin, 1995).

PCR *vs.* Replication

In vitro DNA polymerization is principally quite comparable with the natural replication inside of the cell (Innis, Gelfand, & Sninsky, 1995). Some of the differentiating points are as given:

o   In replication, RNA-based primers are used, while in PCR, DNA-based primers are used designed by using different software.
o   In replication, topoisomerases, helicases and single-celled proteins are used to open up and then to keep double helix apart, while in PCR, this target is achieved by using higher temperature ranges from 91 to 97°C. In replication, DNA polymerase is used, which replicates and proofreads the amplified DNA strand. While in PCR, thermostable polymerase (mostly isolated from *Thermusaquaticus*) is used, which does not have the ability to identify erroneous replication.
o   The mode of replication in both is semiconservative.

Major Chemical Players in PCR

Below is the detail of each chemical constituent of PCR reaction mixture:

DNA Template

Most PCR reactions use DNA as a template molecule to replicate a specific region. The selection of the target site for amplification is selected on the basis of the research question. If it is required to amplify some based region, it is better to convert it into cDNA to enhance its stability in the reaction. This DNA template should be intact to act as a complementary site of annealing of primers and free of inhibitors, which might be the residues left during DNA extraction protocols (van Pelt-Verkuil, Van Belkum, & Hays, 2008).

Oligonucleotide Primers

These are short sequences complementary to the DNA segments that need to be amplified. These primers are designed by using specific bioinformatics software by considering different parameters. The length of the primer determines its specificity to the binding site. The shorter the sequence, the more the chances of reoccurrence of that region in the whole DNA strand. The optimal value for length is usually 17 to 28 base pairs, which provides uniqueness and specificity and is economically manageable. The second factor considered during primer designing is GC content in the overall strand. More the frequency of triple bonds, more energy will be required to anneal and detach primers from the parent strand. More will be the GC content, and more temperature will be required for annealing. Primer's self complementarity and secondary structure formation as dimers or hairpin loops are also avoided during designing.

Deoxynucleotide Triphosphates

dNTPs are used as building blocks for new strand formation. These molecules provide free A, T, C, and G molecules to construct a new phosphodiester bond. dNTPs are attached to the daughter strand in the sequence determined by the parent strand. In the PCR reaction mixture, these dNTPs are used in equal concentration to avoid misincorporation.

Taq Polymerase

It is the only enzyme used in the PCR reaction mixture. This thermostable DNA polymerase is isolated from *Thermus aquaticus* found in hot springs. It has the ability to bear the temperature variations during different stages of PCR. This enzyme binds to the $3^/$ end of primers and extends the daughter strand to the required length.

$MgCl_2$

This bivalent cation is used as a co-factor for the polymerase. Another variant that can be used is Mn which is equally good. This constituent determines the fidelity of PCR reaction and anneals the bases to the extended DNA strand.

Buffer

PCR reaction needs a conducive buffer environment to conduct chemical bond formation. The chemical composition of PCR buffer is 250 mM KCl, 50 mM Tris-HCl pH 8.3, 7.5 mM $MgCl_2$.

PCR Additives

Many additives can be used to speed up the PCR reaction. The most important of them is DMSO (Dimethyl Sulphoxide), which reduces the secondary structure formation. Formamide is used to avoid reannealing of the separated DNA strands. As ancient DNA contains Melanin as a PCR inhibitor, Bovine serum albumin is useful for the removal of such contaminants. Betaine is used to enhance the *Taq polymerase* activity.

PCR Inhibitors

Excess salts, including KCl and NaCl, ionic detergents such as sodium formate, sarkosyl, and SDS, ethanol, isopropanol, and phenol, all reduce PCR efficiency through various inhibition mechanisms.

Different Steps in PCR

PCR is a series of reactions to generate multiple copies of a DNA strand. This process is temperature-dependent. There are three main steps to generate an exact copy of the template DNA strand:

Denaturation

This is the initial step involving opening DNA double helix by breaking the in-between hydrogen bonds. This step requires higher temperature ranges from 91 to 97° C. At this temperature, the DNA strand melted into a single strand.

Annealing

This step involves the attachment of primers to their complementary strand. The temperature of this step depends upon the T_m of the primers. The usual range for this step is 50-65°C. Primers normally wiggle around through Brownian movement and try to fit at the appropriate place. Then H-bond is formed and stays in its place.

Extension

This is the final step conducted at 72°C. At this temperature, *Taq polymerase* binds to the 4$'$ end of the primers and adds more bases by forming phosphodiester bonds.

Exponential Multiplication of DNA

PCR follows the semi-conservation mode of DNA replication. According to this, each strand produces an exact copy of the parent strand after the completion of one cycle. So, existing DNA becomes double at the end of each cycle (Fig. **1**). If 'n' is the total number of cycles in one PCR reaction, then the total number of copies of DNA will be '2n'.

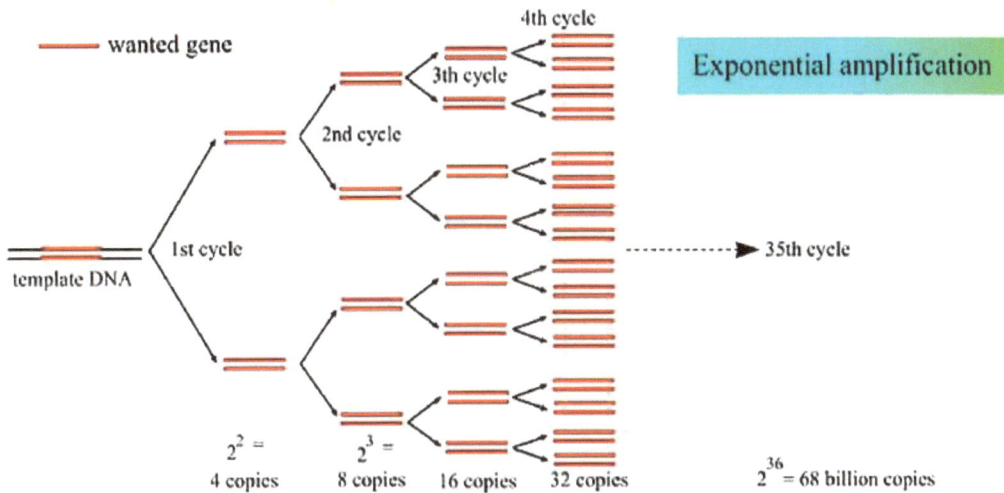

Fig. (3.1). PCR Amplification at an exponential rate.

GENOMIC FINGER PRINTING

In recent years, molecular markers, particularly DNA-based markers, have been widely used in many fields, such as gene mapping and labeling, sex identification, genetic diversity analysis, or genetic correlation. In population genetics, protein-based markers (allozymes) are the first developed and widely used markers. DNA-based methods are now a way to distinguish closely related choices. Furthermore, the use of DNA-based markers allows for efficient comparisons since genetic differences can be detected at all stages of the development of an organism, unlike allozymes that may exhibit age-dependent changes (Dolf, 2013).

The term MARKER is often used for "LOCUS MARKER". Each gene has a specific location on the chromosome called LOCUS. Due to mutations, genes can be mutually exclusive in several forms, called ALLELES (or allelic forms). These markers are functionally important areas that can be used as selection markers for many economically important features. Some popular genomic markers are as follows:

RFLP

Restriction fragment length polymorphism (RFLP) is used to identify mutations at a specific position in the genome. These markers are used as a diagnostic tool to detect deleterious mutations and their existing genotypes in the population. In this method, fragmentation of DNA is done by using restriction endonucleases (see detailed description in Chapter#4). Then fragmented DNA is separated according to length on Agarose gel electrophoresis. And finally, the probe is used to hybridize each fragment. Detection is done on the basis of length variation in different DNA fragments. This method was firstly reported in 1985. But after the completion of the human genome project, the need for RFLP mapping was quite limited. Some drawbacks were observed as it requires a large amount of sample DNA, probe labeling, DNA fragmentation, electrophoresis, blotting, hybridization, washing, and autoradiography. The combined process can take up to a month to complete. But even so far, this method has limited application in marker-assisted selection.

CAPS

It is called Cleaved Amplified Polymorphic Sequence. This method is an extension of RFLP. In this method, PCR-based amplified product is used to target a specific region in the DNA. Restriction endonucleases are used to break the Phosphodiester bond between two adjacent nucleotides from a cleavage site (Fig. **2**).

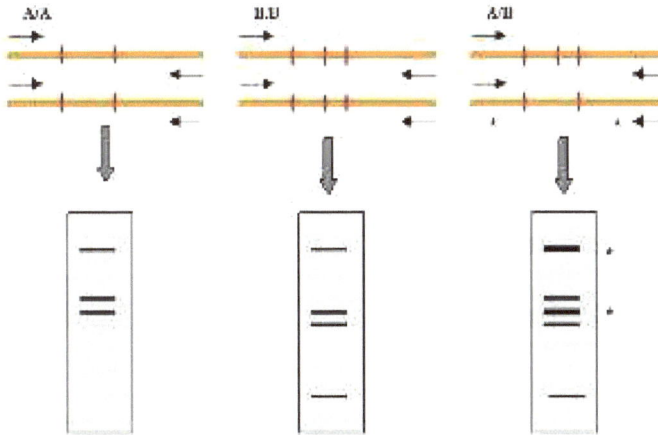

Fig. (3.2). Cleaved Amplified Polymorphic Sequence.

AFLP

It is called Amplified Fragment Length Polymorphism. This is a PCR-based tool for genetics research, DNA fingerprinting, and genetic engineering practice. AFLP was developed by Keygen in the early 1990s to digest genomic DNA using restriction enzymes and then ligated the adaptor to the sticky end of the restriction fragment.

Primers complementary to the adapter are used to amplify a specific set of restricted DNA pieces. These amplified fragments are visualized on denaturing polyacrylamide gel electrophoresis or sequenced by Sanger sequencing. For simultaneous detection of genomic DNA polymorphisms, sensitive and reproducible AFLP technique is used. In different strains and closely related species of taxonomic groups, AFLP has been used widely to find genetic variations. Moreover, it has also been used in forensic, paternity, and quantitative trait loci (QTL).

AFLP technology is capable of simultaneously detecting various polymorphisms in different genomic regions. It is also highly sensitive and repeatable. Therefore, AFLP has been widely used to identify genetic variations of strains or closely related species of plants, fungi, animals, and bacteria. AFLP technology has been used in criminal and paternity testing, as well as in identifying small differences within a population, as well as linkage studies used to generate quantitative trait loci (QTL) analysis maps.

AFLP has many advantages over other labeling techniques, including random amplified polymorphic DNA (RAPD), restriction fragment length polymorphism (RFLP), and microsatellites. Compared to other technologies, AFLP has higher reproducibility, resolution, and sensitivity at the entire genome level and has the ability to amplify 50 to 100 fragments at a time. In addition, no need for prior sequences information (Meudt & Clake 2007) for amplification makes AFLP a more useful tool for taxonomic studies.

Mitochondrial Genome

The mitochondrial genome is also used as a barcode to identify genomically informative regions for species identification and characterization. All eukaryote cells contain mitochondria, and animal mitochondrial DNA (mtDNA) has a relatively fast mutation rate, resulting in the generation of diversity within and between populations over relatively short evolutionary timescales (thousands of generations). The combined effect of higher mutation rates and more rapid sorting of variation usually results in divergence of mtDNA sequences among species and a comparatively small variance within species. A 658-bp region (the Folmer region) of the mitochondrial cytochrome c oxidase subunit I (COI) gene was proposed as a potential 'barcode'. Among some other regions, the displacement loop (D-loop) and Cytochrome-b are also used to identify novel variations to find species differences (Fig. **3.3**).

Fig. (3.3). Human Mitochondrial Genome

SSCP

Single-strand conformation polymorphism is defined as the conformational difference of single-stranded nucleotide sequences of the same length induced by sequence differences under certain experimental conditions. This property allows sequences to be distinguished by gel electrophoresis, and gel electrophoresis separates fragments according to their different conformations (Fig **3.4**).

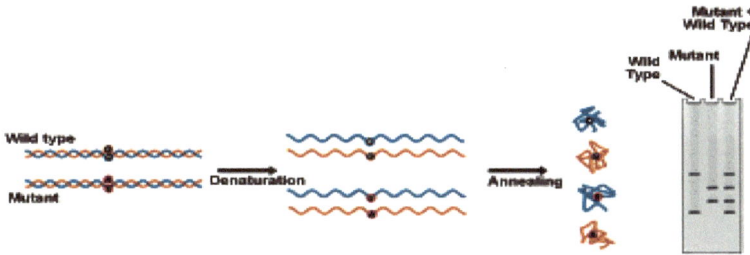

Fig. (3.4). Single-stranded conformation polymorphism

STR

Short tendon repeat sequences are markers commonly used in forensic and genealogical DNA testing. These are repetitive DNA sequences (4 to 8 bases long) typically repeated 5-50 times (Fig **3.5**). Microsatellite analysis became popular in the forensic field in the 1990s. It is used for the genetic fingerprinting of individuals. Microsatellites used today for forensic analysis are all quadruplex or pentanucleotide repeats because they provide highly error-free data while being robust enough to survive non-ideal conditions.

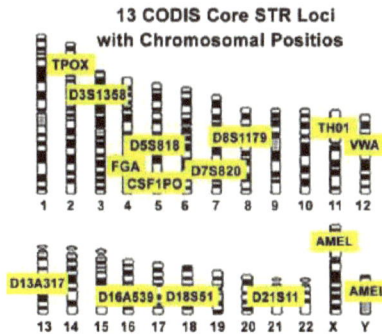

Fig. (3.5). Chromosomal locations of the 13 mini and microsatellite loci in the CODIS panel.

DNA Sequencing

To find the exact sequence of base pairs on a DNA strand was possible after 1974 when two American and English scientist teams independently established protocol. The American team consisted of Maxam and Gilbert, who used "Chemical cleavage protocol", while the English team was led by Sanger, who discovered the "Chain termination method". Interestingly both teams shared Nobel Prize in 1980, but Sanger's method was highly appreciated due to its practicality (Hindley, 2000).

Chemical Degradation Method

This method required radioactively labeled DNA strands, which are chemically degraded at specific sites as Purines (A+G) by using formic acid, guanines, dimethyl sulphate, and Pyrimidines (C+T) by hydrazine and salts for cytosine. The fragments in the four reactions were electrophoretically electrophoresed in a denaturing acrylamide gel for size separation (Fig **3.6**).

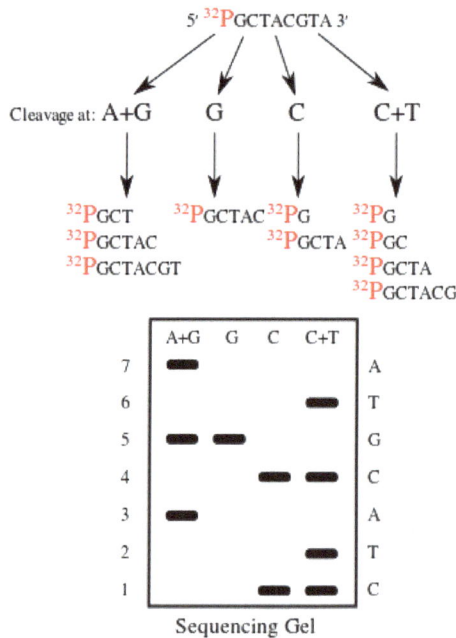

Fig. (3.6). Maxam–Gilbert Chemical Degradation Method.

Chain Termination Method

This method is based on the addition of extra Dideoxy nucleotide bases that lack – OH group at $3'$ atom of ribose sugar (Fig **3.7**). These bases are used in equal proportion with the normal deoxynucleotide bases. Normal PCR reaction causes extension of DNA strand with normal bases, but whenever there is Dideoxy base, the chain is terminated.

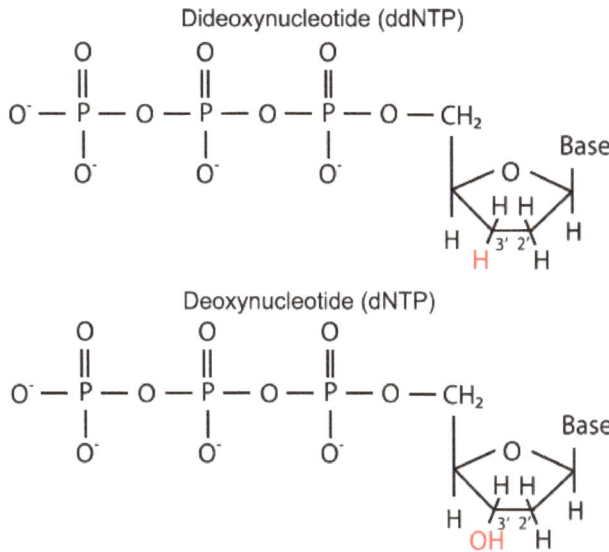

Fig. (3.7). A comparison of ddNTPs and dNTPs.

At the start, this method was conducted manually. The reaction was conducted in four different tubes containing four different ddNTPs; then, it was studied on gel electrophoresis. Now, this method has been automated. All bases are added into the same reaction mixture with a tagging dye of different colors for each nucleotide. The resulted fragments are separated by using capillary electrophoresis. Since four dyes fluoresce at different wavelengths, then these are read on the LASER detector. Results are then illustrated in a chromatogram indicating the sequence and concentration of each nucleotide (Fig **3.8**).

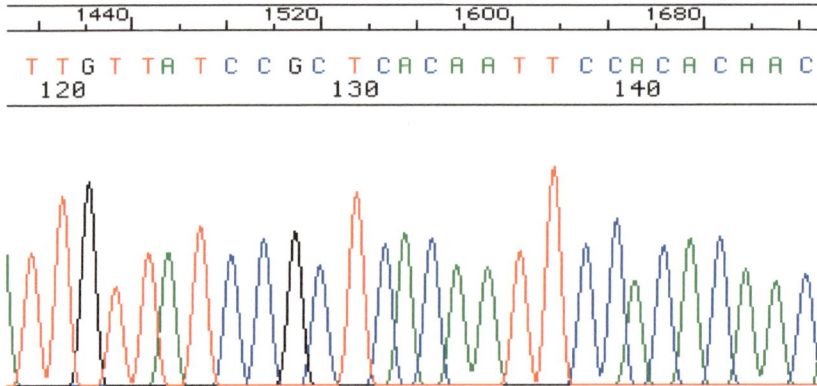

T T G T T A T C C G C T CA CA A T T C CA CA C A A C
  120   130   140

Fig. (3.8). Chromatogram after Sanger's Chain Termination Method.

Pyrosequencing

This method is based on the principle of 'Sequencing by Synthesis'. It relies on the detection of one pyrophosphate release on nucleotide incorporation. This method was firstly developed by Mostafa Ronaghi and Pal Nyren in 1996. This method involves the synthesis of the complementary strand by DNA polymerase with other chemoluminescent enzymes. All four bases are sequentially added, and light is produced after each incorporation. Description of chemical reaction has been given in Fig. (**3.9**) below:

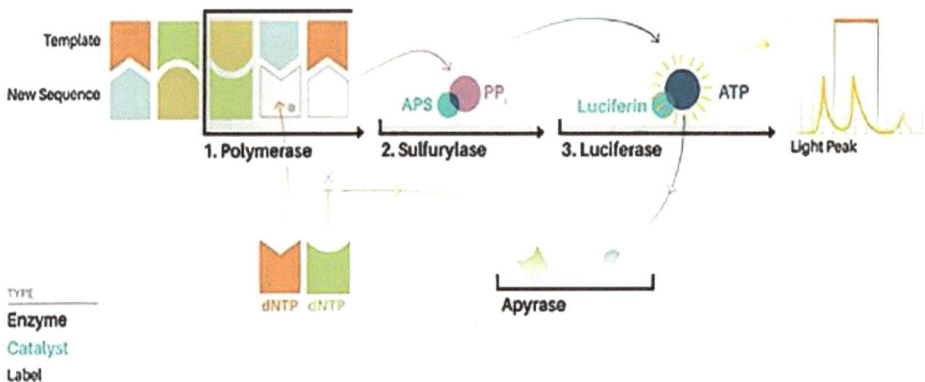

Fig. (3.9). Pyrosequencing.

Next Generation Sequencing

A dire need for low-cost and efficient sequencing at minimal time duration has led to the methods that can sequence 500,000 sequencings by synthesis operations per run (Fig **3.10**). Among many different NGS methods, some are given as under;

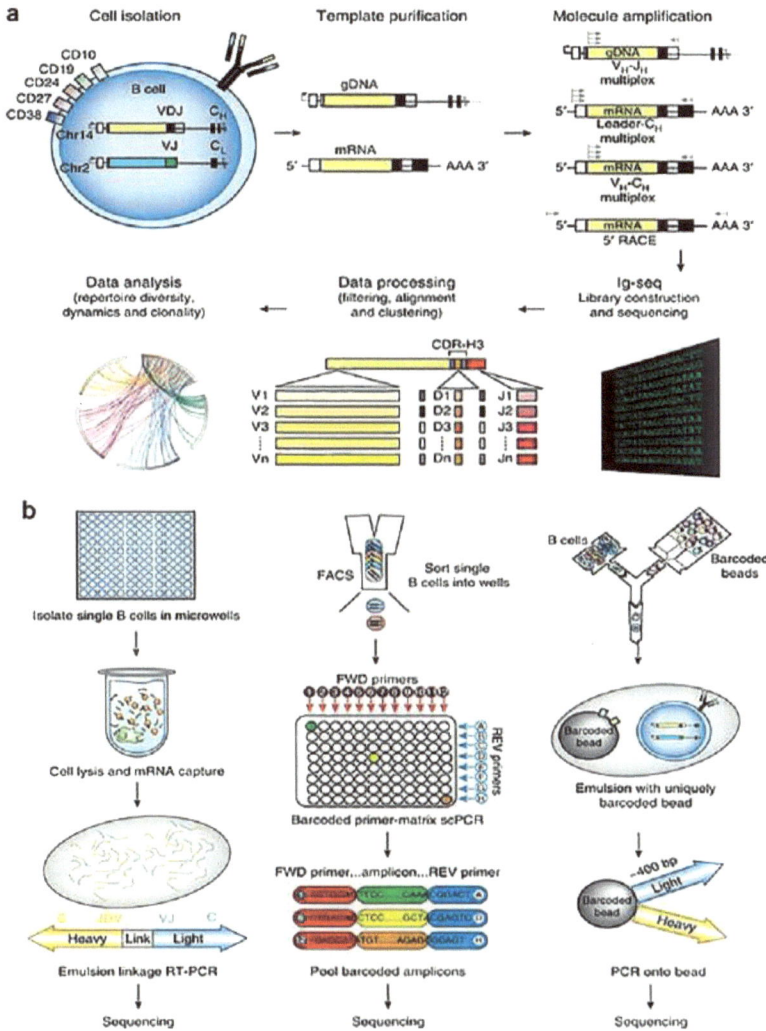

Fig. (3.10). High Throughput DNA Sequencing Methods (www.nature.com)

Polony Sequencing

This sequencing method sequences an E. coli genome with more than 99.99% accuracy. Polony sequencing involves combining *in-vitro* paired tag libraries, emulsion PCR, ligation-based sequencing chemistry, and automated microscopy.

454 Pyrosequencing

Pyrosequencing works on sequencing by synthesis principle where the release of pyrophosphate is detected during a chain reaction which is why it is named as pyrosequencing. The single DNA template is amplified single complementary primer coated on a bead, thus forming a colony of clones. The reaction is performed in a very small well containing a single bead and sequencing enzyme. Luciferase enzyme is used to sense the light emitted upon the addition of a single nucleotide to a growing DNA strand.

Illumina (Solexa) Sequencing

In this type of DNA sequencing, a slide or a flow cell is used to amplify a DNA template with its primer attached to this support (Slide or flow cell), resulting in the generation of a cluster of DNA molecules. Using four types of reversible terminator bases, the sequence is determined. Unincorporated nucleotides are washed away, and the camera captures the fluorescently labeled nucleotides. Before starting a new cycle, terminal 3'end blocker and dye are removed from the DNA chemically. The addition of one nucleotide at a time allows one single camera to capture images of many DNA colonies successively.

SOLiD Sequencing

All possible fixed-length oligonucleotide libraries are labelled according to the sequencing position in this type of sequencing. Annealing of oligonucleotides and ligation through DNA ligase generates the signal information for added nucleotide on that particular site. The amplification of template DNA leads to the production of beads. Each of the beads has one copy of the same DNA. These beads are then deposited on a glass slide.

Ion Torrent Semiconductor Sequencing

The basic principle of ion torrent semiconductor sequencing is the H-ion release and its detection when polymerization of DNA occurs. Very small sequencing wells contain DNA templates and only one type of nucleotides in large quantity to fill the microwell. Complementation of introduced DNA nucleotide to leader template nucleotide results in the incorporation or polymerization of DNA strand. During this incorporation, an H-ion is released. The released H-ion triggers the ion sensor showing the completion of the polymerization reaction.

DNA Nanoball Sequencing

Rolling circle replication amplifies the DNA template into DNA nanospheres. The nucleotide sequence is then determined using strand-free sequencing by ligation.

Helioscope Single Molecule Sequencing

In this reaction, the surface of a flow cell having a poly-A-tail adapter is used for DNA polymerization bases sequencing reaction. The circulating wash of the flow cell contains fluorescently labelled nucleotides similar to that in Sanger di-deoxy sequencing.

SMRT Sequencing

Single molecule real-time sequencing is based on synthetic sequencing methods. Sequencing was performed using an unmodified polymerase (attached to the bottom of the ZMW) and fluorescently labeled nucleotides that flow freely in solution.

Nanopore DNA Sequencing

This method is under development. The growing DNA is passed through a nanopore. Depending upon the shape and size of the DNA, it changes its ionic current. Every type of nucleotide blocks the ion flow through nanopore for different time periods.

Sequencing by Hybridization

In this method, no enzyme is used. The pooled DNA template is hybridized to an array of known sequences.

Microscopy Based Technique

This method relies on direct visualization of DNA molecules in the electron microscope. This allows the identification of an individual base.

Microfluidic Sanger Sequencing

In this method, the entire thermocycling of DNA fragments and its separation by electrophoresis is conducted on a single glass wafer.

CSF Genomics: A Triode

Comparative, structural and functional genomics are interrelated fields in molecular biology to draw fruitful conclusions regarding functional consequences of sequence and structural variations in DNA and proteins, respectively.

Comparative Genomics

This is the field of biological research in which genomic features of different species are compared. These features may include DNA sequence, gene order, and other regulatory regions. This comparison provides the probable sites as functionally significant markers that can be used in animal selection for improved production traits, identification of polymorphisms, conserved regions across species, similar and differential points in genes of one individual or the same gene in different organisms. Most of the outcomes of such analysis provide suitable evidence that justifies major biological evolutionary events in life history (Pagel *et al.*, 2007). Many bioinformatics methods can be used for DNA/RNA/Protein comparison as;

Global Alignment

Global alignments, which attempt to align every residue in every sequence, are most useful when the sequences in the query set are similar and of roughly equal size.

Local Alignment

This method attempts to find the best possible alignment that includes the beginning and end of one or the other sequence (Pevsner, 2015).

Structural Genomic

Three dimensional structure of protein exposes the functionally significant domains that bind and perform a specific task. Proteins are macromolecules that consist of chains of amino acids in a sequence determined by the sequence of bases on DNA strands. Proteins are classified as 'simple' and 'conjugated'. Simple proteins consist exclusively of a polypeptide chain with no additional chemical component added, while conjugated proteins contain one or more non-polypeptide constituents known as 'Prosthetic group'. Some of them are glycoproteins, phosphoproteins, flavoproteins, and metalloproteins (Schmitt, 2008).

Three-dimensional Protein Structure

Three dimensional protein structure indulged four different levels of configuration. The structure mainly includes folds and loops formed by additional non-covalent junctions as Dipole-Dipole, hydrophobic and electrostatic interactions. These configurational modifications are done, after completion of translation, in the step called post-translational modifications. Then final fully folded protein is either secreted out of the cell or consumed inside the cell. Different levels of protein structure are given as under:

Primary Level

This level is defined by a linear array of 20 different types of amino acids and the exact positioning of disulphide bonds. These amino acids are categorized into five classes according to their biochemical nature. These are non-polar (G, A, V, L, I, P) , aromatic (Y, F, W), polar but uncharged (C, S, M, T, N, Q), positively charged (R, K, H) and negatively charged (D, E). These amino acids are bound together by forming the peptide bond between the NH_2 and COOH group after removing one H_2O molecule. This bond is rigid and planer and is in the region of $1.33A^0$ length, so, allows no movements around its plane. Whereas some other bonds, as ϕ and Ψ can move freely rotate around C_α-C and C_α-N at an angle of -180 to +180.

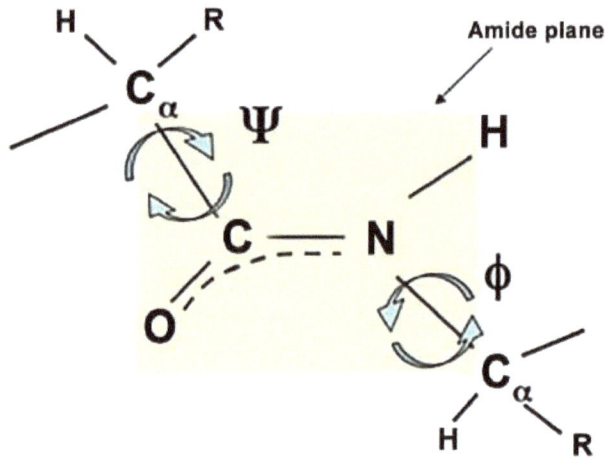

Fig. (3.11). Rotation angels in polypetide backbone.

Secondary Level

The main features of the secondary structure configuration are alpha-helix and beta-sheet. This is the primary level of folding of linear protein molecules. This level is defined as the local spatial conformation of the polypeptide backbone, excluding the side chains that make up the amino acid. The α-helix contains 3.6 amino acids in the entire circle and has a length of about 0.56 nm. The most advantageous amino acids for alpha-helix formation include alanine, leucine, methionine, and glutamine. The occurrence of proline and very close proximity to multiple residues of bulky pendant or pendant groups having the same charge tends to disrupt the formation of the a-helix. The helical structure is stabilized by hydrogen bonding, and each skeleton C=O group forms a hydrogen bond with the N-H group of the four residues preceding the helix. The length of the alpha-helix found in a globular (closely folded, approximately spherical) polypeptide can vary over the length from a single helix to more than 10 consecutive helix turns. The average length is about three turns and is usually located on the surface of the protein. The beta-chain represents the other major repeating structural elements of the protein. The β-strand is typically 5-10 amino acid residues in length, and the residue adopts a nearly fully extended zigzag conformation. A single β-strand is rarely found, but two or more of these chains are themselves aligned to form a β-strand. Suppose all participating β-stretches run in the same direction. In that case, they are called parallel slices, if they run in the opposite direction, they are called anti-parallel, and if the β-folds contain parallel and anti-parallel chains, they are called mixed slices.

Tertiary Level

It refers to the three-dimensional arrangement of all atoms contributing to a polypeptide. The tertiary structure of a small polypeptide (approximately 200 amino acid residues or less) typically forms a single discrete structural unit. However, when examining the three-dimensional structure of many larger polypeptides, the presence of two or more structural subunits within the polypeptide becomes apparent. These are called domains, which are the structural and functional subunits of proteins.

Quaternary Level

This level refers to the overall spatial arrangement of intra-protein polypeptide subunits consisting of two or more polypeptides (Joachimiak, 2011).

Protein Breathing

The constituent atoms of the protein are always moving, and the group from the single amino acid side chain to the entire domain can be replaced by any random motion of any substance up to about 0.2 nm, showing limited flexibility. This movement is called protein breathing.

Computational Methods to Assess 3D Protein Structure

Many computational methods are used to assess 3D structure of the protein as:

Homology Modelling

On the basis of sequence homology of new protein with existing one, its 3D structure can be predicted according to the available existing protein.

Protein Threading

This method is also called fold recognition. In this, instead of the whole sequence, similar folds of proteins are identified, and on this base, the whole structure was predicted.

Ab Initio Modelling

This is also called *De novo* protein modelling. This method builds the protein structure from "scratch", based on physical principles than available structures data (Schwede & Peitsch, 2008).

Functional Genomics

Functional genomics is a new focus of genomic research to elucidate the biological function of these gene products. In the context of genomics, gene function is given a broader meaning, including not only the isolated biological function/activity of the gene product but also:
o   Where in the cell does that product act?
o   How do such influences contribute to the overall physiology of the organism?
Many different methods are used to assign a function to a newly discovered gene, as given under:

Homology Modelling

As mentioned above, this method is used to relate the function of sequence-wise similar genes.

Rosetta Stone Method

This approach depends upon the observation that sometimes two separate polypeptides found in one organism occur in a different organism, a single fused protein, and display a linked function.

Gene Neighborhood

This is another calculation method, depending on the observation that if two genes are consistently side-by-side in several different organisms' genomes, they may be functionally related.

Gene Chips

DNA and protein-based chips are used in Microarray technology to provide information about the expression of functionally active genes by measuring the levels of RNA.

Gene Knockouts

these studies are dependent upon the phenotypic observations after removal of one gene from the living model organism to assess the function it controls (Koonin & Galperin, 2002);(Pevsner, 2015).

Intersecting Triode

Comparative, structural, and functional genomic methods act as a triode and inspecting knowledge the can build conclusions in our understanding of a gene in the living biological system; how it acts alone, how it interacts with molecules around it, how its activity is enhanced or suppressed, what are its positive and negative impacts on the living body, how it binds with its receptors, which pathways it controls or is the part of, *etc.* This information ultimately leads towards the selection of this gene or genetic modification of this gene according to the requirements to get maximum benefits in health, medicine, and enhancing the potential of our animals.

Proteomics

This is the large-scale study of proteins to assess their structural and functional significance as a biological entity. This term was first coined in 1997 to study proteins. These are vital parts of living systems controlling many different metabolic pathways. The proteome is the entire set of proteins in a living cell. As not all genes in a cell are active all the time, so protein synthesis is also demand-driven and varies at different stages of a cell.

This is closely related to functional genomics and requires a systematic and comprehensive analysis of the proteins expressed in cells and their functions. Classical proteomics studies typically require the initial extraction of total protein content from target cells/tissues and subsequent separation of proteins using two-dimensional electrophoresis. The separated protein spots can then be eluted from the electrophoresis gel and further analyzed; mainly for the determination of amino acid sequence. Many different methods as spectroscopy, chromatography, *etc.*, are used to separate and quantify a particular protein from a mixture of as much as about 2000 proteins (Man & Flores, 2012);(Mishra, 2011);(Hardiman, 2009).

Metabolomics

Metabolomics is the study of chemical leftovers of a particular chemical reaction in a cell. This new branch of scientific study depicts the characteristics of saliva or urine metabolites to assess the disease status of the individual. Many advanced methods as gas chromatography, mass spectroscopy, NMR spectroscopy, *etc.*, are used to check the level of different metabolites in biological fluids or in tissues, which can detect the physiological changes due to toxins (Lindon, Nicholson, & Holmes, 2011);(Want, 2007);(Rueda, 2018).

Transcriptomics

This is the study of expression profiling in the given cell/tissue/organism at a specific time. DNA encodes the genetic information into the mRNA that ultimately is translated into protein polypeptides. The need of the cell for a particular protein triggers the gene encoding that protein and results in mRNA formation. Extraction of the total RNA content of a cell determines the production rate of a particular protein in the cell. Outcomes of such studies can be used to predict individuals' health status and assess the functional significance of the production traits in the animals that can be enhanced by triggering the specific genes ON (Wu & Kim, 2016);(Kahl, 2015).

GWAS

GWAS is an attractive approach that has evolved over the past decade into a universal design for investigating the genetic structure of livestock diseases. It revolutionizes livestock genetics, helps detect genes that affect normal variation between species and disease susceptibility, and sheds light on our understanding of economic importance. Researchers working on model organisms are implementing GWAS and other genomic technologies that capture genetic variation.

This is a series of steps to screen and identify the particular disease-associated motif in the genome which may have its role in the trait of interest. In the current era, single nucleotide polymorphism detection in candidate genes is absolute. It is being recommended to emphasize the overall structure of the genome, which may be helpful to identify the different segments or haplotypes that will be involved in different traits of interest like susceptibility/unsusceptibility of disease, morphology, and behavior of the individual's genome to the particular condition. All these corresponding factors fetch the genetics community to reflect on a genome-wide scale (Boopathi, 2013).

CONCLUSION

This chapter presents a detailed description of the principles of DNA polymerization and genomic fingerprinting. It also includes a brief description of DNA sequencing, a triode of comparative, structural, and functional genomics, proteomics; metabolomics; microarrays & transcriptomics. Moreover, rationales for genome-wide association studies are also presented.

REFERENCES

Boopathi, N.M. (2013). *Genetic Mapping and Marker Assisted Selection.* Springer. http://dx.doi.org/10.1007/978-81-322-0958-4
Chin, W.W. (1994). The polymerase chain reaction. In: Kary, B.M., Francois, F., Richard, A.G., James, D.W., (Eds.), *Wiley Online Library* Birkhauser.(p. 458). Boston:
Dolf, G. (2013). *DNA Fingerprinting: Approaches and Applications.* Birkhäuser.
Hardiman, G. (2009). *Microarray Innovations: Technology and Experimentation.* CRC Press. http://dx.doi.org/10.1201/9781420094510
Hindley, J. (2000). *DNA sequencing.* Elsevier.
Innis, M.A., Gelfand, D.H., Sninsky, J.J. (1995). *PCR Strategies.* Elsevier.
Joachimiak, A. (2011). *Structural Genomics, Part A.* Academic Press.
Kahl, G. (2015). *The Dictionary of Genomics, Transcriptomics and Proteomics, 4 volume set.* John Wiley & Sons. http://dx.doi.org/10.1002/9783527678679
Koonin, E.V. & Galperin, M. (2013). *Sequence—Evolution—Function: Computational Approaches in Comparative Genomics.* Springer Science & Business Media.
Lindon, J.C., Nicholson, J.K., Holmes, E. (2011). *The Handbook of Metabonomics and Metabolomics.* Elsevier.

Man, T.K., Flores, R. (2012). *Proteomics: Human Diseases and Protein Functions.* BoD–Books on Demand.
http://dx.doi.org/10.5772/1288

Mishra, N.C. (2011). *Introduction to Proteomics: Principles and Applications.* John Wiley & Sons.

Pagel, P., Strack, N., Oesterheld, M., Stümpflen, V., Frishman, D., Bergman, N.H. (2007). *Comparative Genomics.* Humana Press.

Pevsner, J. (2015). *Bioinformatics and Functional Genomics.* John Wiley & Sons.

Rueda, L. (2018). *Microarray Image and Data Analysis: Theory and Practice.* CRC Press.
http://dx.doi.org/10.1201/9781315215785

Schmitt, L. (2008). *Structural Genomics on Membrane Proteins.* Wiley Online Library.

Schwede, T., Peitsch, M.C. (2008). *Computational Structural Biology: Methods and Applications.* World scientific.
http://dx.doi.org/10.1142/6659

Van Pelt-Verkuil, E., Van Belkum, A., Hays, J.P. (2008). *Principles and Technical Aspects of PCR Amplification.* Springer Science & Business Media.
http://dx.doi.org/10.1007/978-1-4020-6241-4

Want, E. (2007). *Metabolomics: Methods and Protocols.* Wiley Online Library.

Wu, J., Kim, D. (2016). *Transcriptomics and Gene Regulation.* Springer.
http://dx.doi.org/10.1007/978-94-017-7450-5

CHAPTER 4

Gene Manipulation and Genetic Engineering

Abstract: This chapter proposes a brief description of Genetic Modifications of Farm Animals; Strategies for Gene Manipulation; Selection, Screening, and Analysis of Recombination; Transgenic Livestock; Animal Cloning; Biotechnology Applications of rDNA Technology.

Keywords: Cloning, Gene Manipulation, Genetically modified organisms, Recombination.

GENETIC MODIFICATION OF FARM ANIMALS

Farm animals are domesticated for the benefit of humankind. Domestic livestock is used for direct commodities like milk, meat, manure, skin, hide, hooves, horns, eggs, *etc.* Many animals are domesticated for stamina and draught. To get the maximum benefits, animals' genetic potentials are evaluated, and superior traits are selected in breeding programs. But this method is time-consuming, and trait concentration in a certain population takes a much longer time. Recent advances in genetic modification provide the opportunity to change the genome of the organism to get desired traits after single experimentation of genetic engineering. Recombinant animals are also called transgenic organisms, which carry high potentials of production/reproduction *etc.*, and can pass on these traits to the next generations (in most cases). There are many commercially produced transgenic animals and plants. In this chapter, we are going to discuss the strategies to develop a transgenic individual and their examples.

Strategies for Gene Manipulation

Cell structure and function can be studied by isolating a particular component of the cell, but in the case of DNA, it is possible by using relatively small molecules like viral genomes. Even the genomes of the simplest organisms are too large to study directly at the molecular level, and this problem is worsened with genomes of complex organisms. Recombinant DNA technology serves as a milestone to isolate DNA from large genomes. Using the methods introduced by Recombinant DNA technology, gene can be easily purified, its structure can be determined, and

Maryam Javed, Asif Nadeem & Faiz-ul Hassan

the gene's function can be identified by altering it and reintroducing it into cells or whole organisms. Genes can be manipulated by using the following strategies: (Council, 2002a);(Howe, 2007).

Transforming Vectors

Transforming vectors are DNA molecules that serve as a vehicle to carry and transfer genes into cells or organisms where they are replicated and expressed. Typically, vectors should contain an origin of replication, multiple cloning sites, and selectable markers for screening transformants. Some of the major transforming vectors are discussed here (Fig. **4.1**).

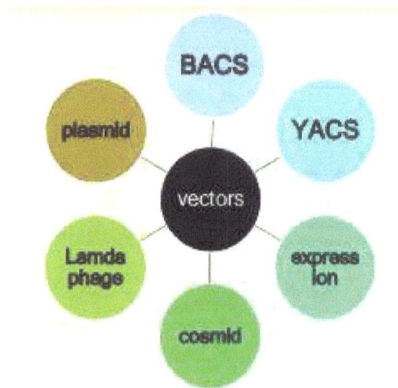

Fig. (4.1). Vectors for Transformation.

Bacteriophage: Bacteriophage is a virus that infects bacteria. The term bacteriophage was devised by D' Herelle, meaning "bacterial eater", explaining the bactericidal activity of viruses. Bacteriophages can be used as vectors. Mostly, a phage contains a linear genome. Two fragments are obtained by introducing a single break in DNA, which can be ligated with foreign DNA. These cloned phage particles are used to infect bacteria, and they can be isolated after the lytic cycle. The limitation of using bacteriophage as a vector is that it cannot take foreign DNA of large size, and this problem is reduced by creating a lambda phage in which non-essential genes have been removed to introduce relatively larger DNA.

Lambda Phage: It is the type of bacteriophage having a head and tail structure. It has high transformation efficiency. DNA is present in the capsid (protein coat), and the tail is the helical structure of the protein. Lambda phage has a temperate life cycle whose lifecycle allows two pathways, lytic and lysogenic. Lambda phage attaches to *E. coli* through its tail and injects its DNA into the host cell. Different phage vectors have been designed for recombinant DNA technology. The genome of lambda phage has specific cos sites, twelve nucleotides sequence with complementary single-stranded ends. Once the phage genome has been injected into the host cell, cos sites of lambda phage circularize. After circularization, the phage genome is either integrated into the bacterial genome to produce progeny or is replicated independently, creating virions. Many phage vectors have been designed to be used in recombinant DNA technology. Lambda phage vectors contain the EcoR1 restriction site and lacZ gene. LacZ gene is important for the screening of recombinant phages. These vectors are used for different purposes like insertion of DNA, selection of recombinants, and expression of genes. Lambda vectors are mostly useful for cDNA or genomic libraries construction. It can take up to 25kb of foreign DNA.

Cosmids: Cosmids are the hybrid of plasmids and phage vectors. Cosmids are named so because they contain cos sites from lambda phage and plasmids from bacteria. Cos sites help in recognition during packaging.

Phagemids: Features of plasmids and phage vectors are combined to form phagemids. A phagemid either contains the bacterial origin of replication and antibiotic-resistant genes for the generation of cloned DNA as conventional plasmids. Some phagemids are composed of phage origin of replication for the production of single-stranded DNA. LacZ gene may also be inserted into phagemids for blue-white colony screening of the transformants. Like cosmids, phagemids can also be packaged into phage particles. In phagemids, phage fragments do not contain the enzymes necessary for replication. So, helper phage particles are used, which provide the machinery important for phagemid replication. Phagemids are important for the production of single-stranded templates for sequencing purposes and the development of templates for site-directed mutagenesis. Long foreign DNA up to 10kb can be cloned.

BAC: Bacterial artificial chromosome vectors are made up of plasmids containing *E.coli* F factor origin of replication and retained a single copy in a cell. BACs were developed for the purpose of cloning and to stably maintain the large fragments of foreign DNA in *E.coli*. They are widely used to study the mammalian genome. Because of their stability, they are mostly being considered for the projects of genome mapping sequencing. Usually, the copy number of bacterial artificial chromosome is maintained at one. It can hold up to 300kb of foreign DNA and involve producing highly stable clones. BAC vectors have been used for whole-genome sequencing of large organisms like humans, mice, rice, and Arabidopsis thaliana. BAC vectors are predominantly used for the development of genomic libraries of complex organisms. BAC also serves as an expression vector for eukaryotic cells. BACs are easy to generate and are important for the development of redundant genomic libraries.

YAC: In yeast artificial chromosome vectors, chromosomes of yeast are ligated into plasmids of bacteria. It can clone DNA fragments of several hundred kilobases in length and is therefore used for cloning of whole genome of an organism. YAC vectors consist of elements like selectable markers, yeast origin of replication, sequences from telomeres and centromere, and some parts of chromosomes for stable segregation, necessary to retain eukaryotic chromosomes in the nucleus of yeast. This vector is generated in a circular plasmid, which is then linearized by restriction digestion. Gene of interest is inserted into the vector and then transformed to obtain desired clones. YACs are mostly used for the expression of eukaryotic proteins that undergo post-translational modifications. This vector is used for the mapping of genomic regions. By using YAC vectors, whole chromosomes can be studied to identify genetic disorders and traits. YACs were used to fill the gaps obtained after sequencing a human genome with the help of BACs.

MAC: Mammalian artificial chromosome vectors are assembled by the combination of three different elements of DNA that include telomeric repeats, replication origins, and centromere. Functions and structures of mammalian chromosomes are complex, and therefore, the preparation of MAC vectors has been relatively slow. MACs can be used for the handling of large genomic regions. They can also be used to analyze the structure and function of genes and gene clusters. Engineering of commercially important livestock is possible by combing MAC technology along with nuclear transfer. Virus-based approaches can potentially be replaced by MAC vectors for the therapies of somatic genes. MACs can be used to retain long foreign DNA permanently and to avoid any unwanted immune response against foreign particles in an extrachromosomal state (Kutter & Sulakvelidze, 2004); (Gussin, 2004).

Host Expression System

The expression system is a genetic construct encoded by DNA and contains the basic machinery for synthesizing RNA and proteins. Advancements in cloning and genetic engineering have enabled the extraction and expression of proteins for experimental purposes. A relatively high amount of protein is required to produce vaccines, antibodies, and enzymes at a large scale. So, protein expression systems should provide a large amount of protein as well as system should be easy to maintain and culture. Prokaryotic and eukaryotic host expression systems are used for the production of recombinant proteins. The culture of host cells, growth rate, post-translational modifications of proteins, and level of target gene expression are some of the parameters for the selection of the right host system. Key elements of the expression system are vector and expression host. Some of the host expression systems are mentioned and discussed in the following section.

Prokaryotic Systems

It is the most commonly used system for the production of recombinant proteins on a laboratory and industrial scale. *E.coli* species is the primary source of this system because of its rapid growth, short replication time. Genetics, genomic sequence, and physiology of *E.coli* are well known. Some of the advantages of the prokaryotic system include the production of a large number of recombinant protein products in a small period of time, mechanisms of transcription and translation of bacteria are well known; bacterial cell cultures are inexpensive, easy to grow and handle, bacterial genomes can be easily modified as well as many bacterial mutants are readily available. Bacterial expression levels can be optimized by altering different parameters. Basic knowledge of the prokaryotic expression system helps in the development of biologically active proteins. The prokaryotic system helps in structural analysis, generation of antibodies, functional assays, and understanding the interaction of proteins. Some of the limitations of this system include the formation of inclusion bodies, poor protein folding in the cytoplasm, codon usage is different in prokaryotes, and minimal post-translational modifications due to which mammalian proteins are difficult to express.

Mammalian Cell Lines

It is important to develop protein in appropriate quality and quantity. Mammalian cells have been progressively used to produce recombinant proteins as they can identify signals for protein synthesis, processing, and secretion. Mammalian expression systems have the ability to induce processes important for a complete biological activity like proper protein folding, product assembly and post-translational modifications. It is a robust, optimized system for a high yield of the protein product. The most common applications of mammalian expression systems include monoclonal antibodies, urokinase production, generation of follicle-stimulating hormones (FSH), expression of complex proteins, and structural analysis. Some of the challenges of mammalian cell lines include the expensive culture media, complex growth requirements, and mammalian cell cultures can be contaminated by animal viruses; therefore, it is difficult to use on an industrial scale.

Insect Cell Lines

Insect cell lines are excellent platforms for producing recombinant proteins vaccines and recombinant antibodies. There are three most commonly used insect expression systems, baculovirus insect cell system (BEVS), *Drosophila* expression system (DES), and insect select (IS) system. BEVS is the most commonly used expression system, especially for transient expression. Baculovirus is a double-stranded DNA, lytic virus, commonly expressed in insect cells of the Lepidoptera family. Its promoters are not active in mammalian cells and it is noninfectious inside vertebrates. Because of its speed and versatility, the BEVS expression system has been used to obtain a high yield of biologically active recombinant proteins like antibodies, scFv fragments, Fab fragments, and other related proteins. Protein processing in insect cell lines is almost similar to mammalian cell lines. Insect expression systems can be used in static or suspension cultures. This system provides good expression levels, especially for intracellular proteins, efficient protein folding, extensive post-translational modifications, and glycosylation like mammalian cells. Limitations of this system include expensive culture media, complex growth requirements, and the production of baculovirus vectors is time-consuming.

Yeast System

Yeast expression system is perfect for the large-scale production of recombinant proteins. It is a well-defined system for the expression of both secretory and intracellular proteins. Yeast is an attractive organism for expression system due to its highly developed genetic system, ease of use, low cost, ability to take recombinant plasmids, and yeast cultures can be grown to very high densities. This system is suitable for the production of recombinant proteins at a large scale. The most commonly used yeast strains for use in host expression systems are Saccharomyces cerevisiae and Pichia pastoris. The main advantages of the yeast expression system include high yield, chemically defined medium, product processing similar to mammalian cells, stable production strains, durability, and lower protein production costs.

Some challenges of this system are growth conditions that may require optimization, and fermentation is required for very high yields.

Plant System: Plants are the source of essential biological molecules. Plants can be used as an expression system for the development of recombinant proteins. As compared to prokaryotic and mammalian expression systems, protein development in plants is safe and economical. To use plants as an expression system, identification of such promoters is important, associated with high yield and enhancing the stability and recovery of protein. Proteins obtained from the plant expression system are better in terms of storage and distribution. Plant expression systems have a minimum risk of being contaminated by microbes, pathogenic to humans. Plant cultivation is relatively less complex. Plant cells like seeds or tubers can be used to express a protein of interest, where proteins can be transported easily and stably. The challenges of the plant expression system are that plant proteins do not have terminal galactose, mostly found in animals. Post-translational modifications of plants slightly differ from animals (Old & Primrose, 1981);(Satyanarayana & Chakrapani, 2013);(Fernandez & Hoeffler, 1998).

Molecular Scissors

Molecular scissors are the restriction enzymes used to cut the DNA at or near specific sites called restriction sites. These enzymes were first discovered in bac-*Identification of the clone* teria and got importance because of their ability to inhibit phage particles' growth. Viruses inject their DNA into bacteria and attack the host replication machinery to produce their progeny. Bacterial cells have developed a protective mechanism in which they use restriction enzymes that destroy viral DNA by cleaving its specific restriction sites. Restriction enzymes break the phosphodiester bond by hydrolyzing the bond between 3' oxygen and phosphorus atoms (Fig. **4.2**). The enzymes are catalyzed by divalent ions. Some restriction enzymes cut at both strands of DNA at the same position, thus creating blunt ends.

Other restriction enzymes cut at a different position on both strands producing single-stranded overhangs at the ends and known as sticky ends.

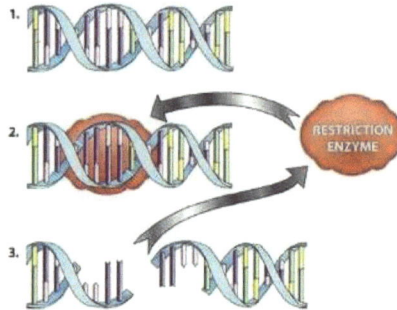

Fig. (4.2). Restriction Enzyme.

Types of Restriction Enzymes

Two major classes of restriction enzymes are endonucleases, which cut within the DNA sequences, and exonucleases, which cut at the ends of sequences. Restriction enzymes are categorized into four classes on the basis of their composition, restriction site, and co-factor requirements.

I.   Type 1: This type of restriction enzymes cut at sites away from recognition sites, and they have a restriction as well as methylase activities. Cofactors of type 1 are ATP, AdoMet and Mg^{2+}. Examples include EcoB and EcoK.

II. Type 2: Enzymes in this category either cleave within or at a short distance from the recognition site. The cofactor required for type 2 is Mg^{2+}. EcoR1 and BamH1 are examples of type 2.

III. Type 3: Type 3 enzymes cut at 25 to 27 base pairs from the recognition site. ATP and Mg^{2+} are required cofactors. EcoP1 and Hinf III are examples of type 3.

IV. Type 4: Not much is known about type 4. These enzymes cut close to or within the recognition site of the methylated DNA. Co-factor required for type 4 is Mg^{2+}.

Restriction enzymes are widely used in molecular cloning, DNA mapping, restriction landmark genomic scanning, sequencing of genes, restriction fragment length polymorphism (RFLP), pulse-field gel electrophoresis, serial analysis of gene expression (SAGE), and restriction enzyme-mediated integration (REMI).

Selection Screening and Analysis of Recombinants

It is important to identify the transformed host cells once the recombinant DNA has been introduced into the host cells. Different methods are available to differentiate the transformed cells from recombinant cells, and these methods can be categorized into two basic concepts.

Direct Selection: It is a quick and comprehensive method of choice. To select the desired cloned gene, it is essential to culture the transformants on the agar plate, on which only the recombinants containing the gene of interest will grow. Most vectors are designed so that insertional inactivation of the genes present on the molecules will occur after the introduction of the gene of interest. However, the direct selection method does not apply to all genes. The two most common examples of direct selection methods are discussed here.

Direct Antibiotic Resistance Screening: Usually, antibiotic resistance genes are inserted into the vectors. These genes are responsible for the destruction of the integrity of one gene present in the molecule. For example, if the ampicillin resistance gene (amp^r) is introduced into the vector, it will introduce ampicillin resistance in the transformants. It means that when these clones are cultured on agar plates containing ampicillin antibiotics, only cells containing antibiotic-resistant markers will be able to grow on culture media. This method is helpful only to differentiate the recombinant cells from non-recombinant cells. And thus, it cannot help in screening recombinants containing the gene of interest or the recombinants that are religated vectors without having any gene of interest.

Blue White Screening Method: It is a more sophisticated method of screening. Insertional inactivation of the gene also occurs in this method. LacZ gene is introduced in the vector. This gene encodes the β-galactosidase enzyme. If expressed as blue color colonies, this enzyme hydrolyzes a chromogenic substrate called X-gal on a culture plate. When the gene of interest is inserted into the lacZ gene, it will disrupt the gene's activity, and the β-galactosidase enzyme will not be encoded by it so that white-colored colonies will form.

Identification of the Clone From Gene Library: Different methods are available to identify the clones from genomic libraries. Some of these techniques include selection by functional analysis and nucleic acid hybridization. The functional screening method involves the use of an expression vector to express the protein of interest. In this method, an expression library has been established that allows the functional expression of cloned fragments. These desired cloned fragments can be identified by immunological screening (antibodies), by utilizing the known activity of gene product, or by some other ligands which specifically identify the encoded proteins. Nucleic acid hybridization plays an essential role in the identification of individual clones in a library. In this method, labelled probes are used for the detection of complementary sequences. DNA is transferred onto a nylon membrane. Treat the membrane with alkali to produce single strands of DNA and provide heat treatment or UV irradiation to bind the strands on the nylon membrane. Then hybridize the membrane to its complementary probe in the solution having radioactively labelled probes. Wash the membrane to remove unattached probes and visualize the hybridized probes with autoradiography (Nicholl, 2008);(Wu, 2012).

Transgenic Livestock – An Emerging Technology

Since the beginning of animals domestication, animal biotechnology has been in use. Transgenics is an emerging technology, and it allows the rapid introduction of new and modified genes or fragments of DNA into livestock without hybridizing or crossbreeding. It is a complicated technique, but its results are similar to those of genetic selection and crossbreeding. The development of transgenic livestock will significantly help in the improvement of nutrition, health, protection of the environment, animal welfare, and protection of livestock from several diseases (Niemann & Wrenzycki, 2018).

Herman: The Bull

Herman was the first genetically modified bull in the world. It was genetically modified at the early embryo stage. It was injected with the human gene encoding for lactoferrin. Cells were then cultured *in vitro* to the embryo stage, and then these cells were transferred to recipient cattle. In 1994, Herman became the father of eight calves, and all of these calves carried the genes encoding for lactoferrin. Herman was euthanized on April 2, 2004, because it was suffering from osteoarthritis.

Salmon: The Fish

Salmon are slow growers and only gain weight between 20 to 30 grams in the first year of their life. To overcome this problem, transgenic salmon have been developed by isolating the growth hormone of Chinook salmon and introducing it into fertilized Atlantic salmon eggs. This transformation allowed the salmon to gain more weight in less time. Transgenic salmon also has the ability to grow in hot water, domestically in close containment populations.

Biosteel Goat

Biosteel goat is a genetically modified goat that can produce golden orb weaver spider silk protein in its milk. Recombinant proteins obtained from goats are subjected to chromatographic techniques for the purpose of purification. Spider silk-like protein obtained from transgenic goats is 7 to 10 times stronger and can be stretched up to 20 times without losing any of its strength, and it is resistant to highly extreme temperatures ranging from -20 to 330 °C.

Transgenic Goats

The gene of interest has genetically modified transgenic goats to introduce the desired traits. Transgenic goats can be developed to improve goat breeds, for high milk yield, to improve the quality and quantity of meat, and for better nutrition purposes.

Animal Cloning

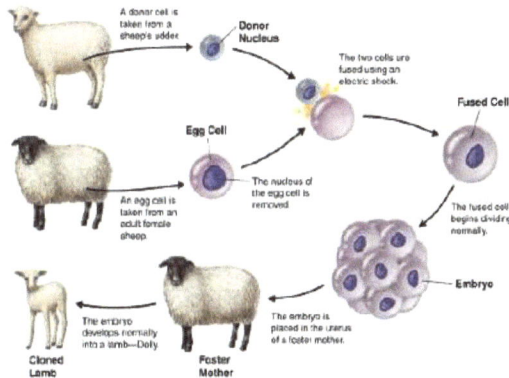

Fig. (4.3). Animal Cloning.

It is a kind of assisted selective breeding of animals. Animal cloning is a complex procedure in which a complete organism is produced from the single-cell copied from the donor animal. Cloned animals develop the exact duplicate of the donor. It is a reliable way of producing genetically superior animals. The main reason to clone the animals is to gain such products that are beneficial to humans. For instance, scientists have introduced human genes in cows or sheep to obtain blood clotting factor IX to treat hemophilia, cystic fibrosis, and lung problems. Cloning helps to increase the production of productive animals to gain better and healthy food. Cloning minimizes the use of growth hormones, antibiotics, and other chemicals by developing the healthiest animals. Endangered species can also be protected by cloning (Council, 2002a).

Reproductive Cloning

Reproductive cloning is the development of genetically identical animals *via* somatic cell's nuclear transfer from a donor organism. In this method, nucleus of the somatic cell is removed from the cell (Fig. **4.3**). At the same time, the nucleus of the egg is also removed. Nucleus removed from a somatic cell is incorporated into the enucleated egg cell. The egg containing somatic cell nucleus is subjected to division with the help of a shock. After several divisions, a blastocyst is formed, having DNA almost similar to the donor organism. The newly formed embryo is implanted into the uterus, where it develops. Dolly sheep is an example of somatic cell nuclear transfer. It has been widely used for mass production of animals, interspecies nuclear transfer, stem cell research, regenerative medicines, conservation of wild animals for next generations, xenotransplantation and development of disease-resistant animals,

Dolly

Dolly is the first mammal to clone from adult cells by somatic cell nuclear transfer successfully. It was born on July 5, 1996, and its original code name is "6LL3".

It was cloned and kept for her entire life in Roslin Institute in Midlothian, Scotland. Dolly had three mothers, one was the egg donor, and the other provided the nucleus from mammary gland cells, whereas cloned embryo was implanted in the third one. Dolly gave birth to a total of six lambs. At the age of four, it developed arthritis and died on February 14, 2003. Production of Dolly proved to be revolutionary for the cloning of other large mammals like bulls, horses, and deer.

Molly

Molly was the first mammal to be successfully cloned and become transgenic at the same time. It was born in 1997 at Roslin institute. It was developed by using the process of somatic cell nuclear transfer in which a new gene was introduced, which was responsible for the production of a therapeutic protein, human blood clotting factor IX, in its blood. Fibroblast cells were used as the source of the nucleus. Production of Molly gave insight into the therapeutic potentials of recombinant DNA technology when combined with animal cloning (Panno, 2014).

Biotechnological Applications of rDNA Technology

Recombinant DNA technology is a sophisticated branch of molecular biology established to develop biological products on a large scale to fulfill the demands of the From research and biotechnology to the medicines stocked on the shelves of pharmacies, rDNA technology has many applications (Verma & Trounson, 2006). Various applications of rDNA technology in biotechnology are out listed below.

Applications in the Improvement of Crops

Recombinant DNA technology has several potential applications for the improvements of plants; some are given here.

Distant Hybridization

It is the transfer of genes between distantly related species. With the progress of genetic engineering and recombinant DNA technology, the gene of interest can even be transferred from lower to higher organisms.

Transgenic Plants

Plants that have been genetically modified and contain foreign genes are known as transgenic plants. Tolerance to metal toxicity, being disease-resistant, insect and pests resistant, herbicides resistant and drought-resistant plants, male sterility for plant breeding purposes, and quality improvement can be gained through recombinant DNA technology. For example, BT cotton is resistant to bollworms.

C4 Plants

C4 plants have been developed to improve the yield of plants. The photosynthetic efficiency of plants can be improved by converting C3 plants to C4 with the help of protoplasm fusion or by recombinant DNA technology. C4 plants are better than C3 plants as they have a higher rate of biomass production. Tropical and subtropical zones are better for the growth of most of the C4 plants. Sorghum, sugarcane, and maize are some common examples of C4 plants (Roller & Harlander, 1998); (Congress, 1992).

Applications in Medical Diagnosis of Disease

Biotechnology, especially recombinant DNA technology, serves as a tool to diagnose several diseases. For example, genetic testing can be done to diagnose genetic disorders. Similarly, probes can be used to identify infectious agents like hepatitis virus or HIV. Genetic engineering plays a key role in producing vaccines, hormones, and antibiotics.

Antibiotics

Recombinant DNA technology is used for the mass production of antibiotics. Fungi like Penicillium and Streptomyces are used for the development of famous antibiotics penicillin and streptomycin, respectively.

Vaccines

Vaccines can be developed by transferring the genes encoding antigen into the disease-causing bacteria. These antibodies prepare the body against the infection of similar bacteria or viruses.

Interferons

Interferons are antiviral proteins produced as a result of viral infection. They serve as the first line of defense against viral attacks. Naturally, they are produced from the blood cells of humans in very small amounts, so they are very expensive. Recombinant DNA technology has made it possible to develop interferons in large quantities and much cheaper in rate.

Gene Therapy

Gene therapy replaces a defective gene (responsible for hereditary diseases) with a normal and healthy gene. It can be used to treat several diseases like sickle cell anemia, Severe Combined Immunodeficiency Syndrome (SCID), hemophilia, and phenylketonuria. The success rate of gene therapy depends on the development of a better vector that can be sustained for the long-term expression of newly induced genes and on the development of a better understanding of the physiology of genes (Council, 2002b).

Cloning

The discovery of cloning has revolutionized the direction of research. Some of the cloning types are given here:

Microbial Cloning

When the microbial cells have been genetically altered, they are cultured on a growth medium, and millions of clones can be obtained within a few days. Production of human insulin from *E.coli* and Bt toxin from *B. thuringiensis* are common applications of microbial cloning.

Animal Cloning

Animal cloning plays a vital for the production of superior animals. It helps to get better milk yield, high-quality meat, increase production of better breeds, and gain therapeutic products.

Plant Cloning

Root and shoot tips of plants are widely used for cloning purposes. It is used to increase the production of important crop plants and horticulturists in a short time. Flavr save tomato, and golden rice are examples of plant cloning.

Industrial Applications

Recombinant DNA technology can help produce efficient microbial strains for the generation of commercially important chemical compounds, development of proteins from some waste products, and the improvement of fermentation procedures in industries. Microbes can also be used to clean up pollutants from the environment.

REFERENCES

Congress, U.S. (1992). *A New Technological Era for American Agriculture.* ITA-F-474. Office of Technology Assessment, Washington, DC.

Council, N.R. (2002b). *Scientific and medical aspects of human reproductive cloning.* National Academies Press.

Fernandez, J.M., Hoeffler, J.P. (1998). *Gene expression systems: using nature for the art of expression.* Elsevier.

Gussin, G. (2004). A Genetic Switch-Phage Lambda Revisited By Mark Ptashne. *BioEssays, 26*(11). http://dx.doi.org/10.1002/bies.20148

Howe, C. (2007). *Gene cloning and manipulation.* Cambridge University Press. http://dx.doi.org/10.1017/CBO9780511807343

Kutter, E., Sulakvelidze, A. (2004). *Bacteriophages: biology and applications.* Crc press. http://dx.doi.org/10.1201/9780203491751

National Research Council. (2002). *Animal Biotechnology: Science-Based Concerns.* National Academies Press.

Nicholl, D.S. (2008). *An introduction to genetic engineering.* Cambridge University Press. http://dx.doi.org/10.1017/CBO9780511800986

Niemann, H., Wrenzycki, C. (2018). *Emerging Breeding Technologies.* Springer. http://dx.doi.org/10.1007/978-3-319-92348-2

Old, R.W., Primrose, S.B. (1981). *Principles of gene manipulation: an introduction to genetic engineering.* Univ of California Press.

Panno, J. (2014). *Animal cloning: the science of nuclear transfer.* Infobase Publishing.

Roller, S., Harlander, S.K. (1998). *Genetic Modification in the Food Industry: A Strategy for Food Quality Improvement.* Springer. http://dx.doi.org/10.1007/978-1-4615-5815-6

Satyanarayana, U., Chakrapani, U. (2013). Biochemistry, Book and Allied Pvt. Ltd. *Kolkata.*

Verma, P. J., & Trounson, A. (Eds.). (2006). Nuclear transfer protocols: cell reprogramming and transgenesis.

Wu, R. (2012). *Recombinant DNA Methodology II.* Academic Press.

CHAPTER 5

Genes *Vs.* Environment

Abstract: This chapter proposes a brief description of Epigenetics; Role of DNA, RNA and Chromatin; Epigenetic regulation of Chromosomal Inheritances; Structural & Biochemical Advances in Mammalian DNA Methylation; Histones Modifications; Epigenomics.

Keywords: DNA Methylation, Epigenetics, Epigenomics.

EPIGENETICS

Epigenetics means "top" or "above" of genetics. It relates to the inheritable change in phenotype by modifying gene expression (active *versus* inactive) rather than altering the genetic sequence. These all variations are inheritable and last throughout cell multiplication for the period of the cell's life cycle, even remaining for many generations without altering the DNA sequence. Histone modification and DNA methylation are two types of changes in a chromosome that can alter gene expression without altering the DNA sequence. The activity of the repressor protein can also control gene expression. Epigenetic changes are stimulated by external environmental factors or may be part of the natural developmental process. The cellular differentiation process in eukaryotes is an example of epigenetic change. During morphogenesis, the zygote divides, producing daughter cells that differentiate into all the diverse cell types of an organism, including skin cells, neurons, liver cells, muscle cells, epithelium, and blood vessels endothelium, *etc.* activating some genes while inhibiting the expression of others. Epigenetic modification can have a lot of damaging effects that can be directed to diseases like cancer. A variety of disorders occur due to the epigenetic component responsible for the unusual regulation of a gene. For instance, blood cell disorders represent the best example of this kind of unusual gene regulation. Many components regulate the process of differentiation, activation, and growth of blood cells. When these blood cells are arrested in unusual regulation of genes, the effect can be inflammatory disorders, unusual growth, auto-immune and leukemia, *etc.* (Blazer & Hernandez, 2006).

Joseph Conrad H. Waddington initially outlined the term epigenetics in 1942, derived from the Greek word "epigenesis" that formerly related to the differentiation of the cell from its state of totipotent. He used it as an abstract model of how genes produce a phenotype by associating with their surroundings. Robin Holliday studied DNA methylation and cellular memory, reported that methylation of DNA sequences has great influences on the expression of genes and influences continued throughout cell division. Holliday proposed two definitions of epigenetics, when taken separately, which were admittedly inadequate but satisfied all presently recognized processes of epigenetic once taken simultaneously. The primary definition was "the study of the variations in the expression of a gene, that happen in living bodies within cells which have differentiated, and also the mitotic inheritance of these patterns of expression." Another definition formulated is that "genome inheritance, doesn't depend on variations in DNA." Morris and Wu (2001) interpreted the definition of Holliday and stated: "the study of modification in function of a gene are meiotically or/and mitotically inheritable which don't require variation in DNA." The inclusion of inheritability in the Waddington's by Holliday was a major meaningful modification. At the same time, the definition of Waddington doesn't exclude inheritance of expression states.

It is still less strange to understand why some genes are switched off or on; the epigenetic field started due to the characterization of regulatory gene-protein and gene-gene associations. These conclusions justify the modification in the expression of the gene that Waddington called epigenetics. However, the actual problem satisfies the addition of inheritability in Holliday's definition. DNA encodes all these Regulatory components, but Holliday's epigenetic theory states that the expression, of these components is not only required for expression of the gene, but is inheritable. This phenomenon further needs a new mode of inheritance that is independent of DNA sequence. To fully understand the definition of Holliday, each element should be accurately outlined. This needs a serious aspect of the interpretation as Holliday outlined the terms inheritability, dependence, and DNA sequence (Armstrong, 2013).

The term dependence provides many possible meanings. Strictly, anything which can't come into existence without DNA may be thought of to be dependent of DNA. Consequently, anything which depends on the genomic sequence for its production, activation and/or maintenance is dependent, and it might add anything which needs DNA sequence as a substrate. According to this point of view, anything from DNA methyltransferase encoded by particular genes DMNT to histones that utilize DNA as a substrate throughout alteration can be recognized as DNA dependent. It is expected that Holliday may claim that it is usually not the purpose he had in mind when he made this difference. Rather, he relates to dependence in a highly strict way as the association among stable expression state, the position of a specific chromosomal locus, and a particular nucleotide sequence of DNA inside this locus. For instance, according to arguments of Holliday, a capability of a similar sequence of DNA to represent a totally distinct phenotype without a nucleotide modification shows freedom on the initial sequence of DNA because there must be something out of a sequence of DNA that is regulating the gene expression. This then needs to be understood as what accurately is meant by DNA sequence. Several techniques of molecular biology are employed to study further the phenomena of epigenetic phenomena, including DNA adenine methyltransferase identification (DamID), fluorescent *in situ* hybridization, bisulfite sequencing, immunoprecipitation of chromatin, and methylation-sensitive restriction enzymes (Huang, Litt, & Blakey, 2015).

Role of DNA, RNA, and Chromatin

Many characteristics of DNA sequence are usually ignored. Almost all geneticists are mainly concerned with the euchromatic part of the chromosome that contains protein-encoding genes. Unsurprisingly these regions of chromosomes effectively generate almost all protein types necessary for the function of cells and survival. Repetitive sequences and those found within the heterochromatin are usually ignored and commonly referred to as junk DNA. Repetitive sequences may be ignored because their roles are less understood, and the techniques for investigating them are not fully developed. Our concern with the region which encodes the protein and the problem in studying the part of repetitive sequences has formed, made restricted, our belief on the gene sequence plays a role in its expression. But experiments show that besides nucleotide sequence, there are other features of a genome necessary for the expression of the gene. It may depend on other DNA sequences that exist out of the region that encodes protein. These conclusions reject the association between the expression of a gene and its primary sequence. Gene expression of a particular gene may depend on another DNA sequence. These arguments can be fixed by extending the definition of a gene by adding control regions and the precise requirement to outline the locus of regulative modification. Including a regulatory region in a gene cause uncertainty and problem in accurately defining the range of genes (Maclean, 1986).

The second characteristic of DNA sequence that is usually ignored is the location of sequence, which might influence the expression of the gene in non-coding and coding parts. The movement of a DNA sequence to another position in the genome may affect its expression, and in these circumstances, independence is continued to be supported by many epigeneticists till the alterations happened in the transposed sequence. Transgenesis, general observation in biology, is very obvious that the inserted position of the transgene has a meaningful influence on the expression.

Copy number of nearby sequences is the third critical aspect of the gene sequence. Repetitive regions play essential regulatory roles. The proximity of coding regions to repeat may also include the size of repeating regions and have noticeable effects on gene expression and structure of chromatin.

RNA function and their association with proteins and DNA can alter gene expression; therefore, more studies are now focusing on non-coding RNA as epigenetic players to study their role in gene expression.

Particularly, increasing emphasis is being placed on the capability of non-coding RNA (ncRNA) copies to change gene expression and, thus, their role as epigenetic modifiers. Environmentally critical, efficient regulation of gene expression by long and short ncRNAs results in indirect and direct association with classical epigenetic mechanisms, developing a large epigenetic network. ncRNAs bind and recruit histone-modifying complexes to either remove or add acetyl and methyl groups. These ncRNA further modulate DNA methyltransferases by suppressing or facilitating DNA methylation. Loci that are targeted by each of these mechanisms can encode non-coding RNA and/or protein-coding transcripts. ncRNAs, in turn, can change gene expression by associating with mRNA.

Several molecules called chromatin remodelers offer the mechanism for remodeling the chromatin and providing transcription signals to reach their targets on the DNA strand. Chromatin remodelers are large, multi-protein networks that use the energy of ATP hydrolysis to assemble and restructure the nucleosomes. Nucleosomes consist of approximately 1.7 turns of 146 base pairs of DNA around a histone-octamer disk, and the DNA within each nucleosome is frequently unavailable to DNA-binding factors. Remodelers are thus essential to afford access to the DNA to allow transcription, chromatin assemblage, DNA repair, and other processes. A nucleosome core contains eight histone core proteins (two each of H4, H2A, H3, and H2B) and 146 nucleotides of double-stranded DNA. Canonical histones can themselves be modified by specific enzymes or substituted by histone alternatives, thereby shaping the neighboring DNA more or less available to the transcriptional assemblage. For example, H2A.Z is an alternative to H2A and is often enhanced near relatively inactive gene promoters. Surprisingly, H2A.Z does not hold its position during DNA replication when the chromatin structure is stabilized. Instead, the chromatin remodeling complex SWR1 catalyzes an ATP-dependent exchange of H2A in the nucleosome for H2A.Z. CENP-A is another known histone alternative that has been discovered to be linked with centromeres. Originally restricted to the centromere through immunofluorescence studies, CENP-A was linked to centromeric activity through cell division. But, once the CENP-A protein was isolated and sequenced, it was shown to have sequence homology to H3, proposing that CENP-A truly replaces canonical H3 near the centromere. Some experiments propose that these alternative histones that happen in selective areas of the genome may aid in the specific regulation of chromatin role and gene transcription from these regions.

Epigenetic Regulation of Chromosomal Inheritance

Currently, there is controversy in the field as to how permanent and inheritable a mechanism must be in order to be classified as "epigenetic" and whether it is admissible to hold not only those marks maintained over generations of organisms but also those transported during cell divisions, simultaneously with other kinds of self-preserving marks. Epigenetic marks associated with gene suppression and patterns of DNA methylation have been shown in both plants and mammals to be stably movable to offspring. On the other hand, DNA methylation status is not transferred with perfect precision, making alterations in the pattern even, between clonal cell populations. Moreover, although histone protein modifications are recognized as general epigenetic mechanisms, they are dynamically modified by demethylases, and their distribution can turn over within hours in response to environmental stimuli (Meyers, 2012).

Inheritability is the most significant and definitive component of epigenetics. The definition of epigenetics was simultaneously constricted and expanded by the addition of inheritability. Including inheritability into the analysis insist on studying epigenetics by considering the association among expressional changes and stimulus which is reason to make permanent or temporary influence on that modification, expecting the alteration in expression to proceed by meiosis or/and mitosis in order for a phenotype to be considered epigenetic extremely reduces the number of measurements which qualify. Because of these reasons, the inheritability features of the definition explained by Holliday are uncertain, especially because there is need to get to know the new method of inheritance. Definition prospect, the inclusion of inheritability extends the sense of the term epigenetic because inheritability is associated with the transformation of just DNA. Inheritability explains the transformation of molecules other than DNA; many cytoplasmic components, histones, or methyl groups widen the sense of inheritance. But, definitions of Holliday do not truly describe the distinction between the inheritance and transfer of molecules. Also, it does not explain which kind of molecules will transfer which will not. Without difference between them, it becomes problematic to distinguish between non-epigenetic phenomenon from epigenetic, and the mode of inheritance may also be unknown. Holliday's sense of inheritability creates many complexities. First, it isn't easy to distinguish among alterations in the gene expression because of the inheritance's state of expression and those that are because of immediate feedback to a stimulus. Before explaining the state of expression that's inherited, one must be clear about knowledge of stimulus. Understanding association among effects of expression and stimulus effectively produces time-line and conclusively expresses restriction present among both for which inheritance is significant. Specific expression pattern is produced in the parent organism because of a particular stimulus, and this pattern of expression is obvious in the daughter as well without facing that stimulus. Whereas these associations are enough to imagine, they can be challenging to demonstrate practically, not only due to the changeable expression of the gene, but also because several states' stimuli influencing a parental organism may also affect the germ cells present in the parental organism, germ cells that generate offspring. Suppose germ cells react to stimuli encountered by the parental organism. In that case, no restriction is present among the stimuli and daughter because primordial cells show that the expected daughter's expression is directly affected as well. For instance, in vertebrates, any stimulus influencing pregnant organism-bearing offspring can affect fetus, mother, and the germinal cells present in the daughter, generating further progeny. This indicates that stimuli faced by a female can directly affect further possible progeny. If the pattern of expression in primary germ cells were obvious in the daughter, it would, however, fulfill the definition of Holliday as a preserve which mitotically would have had to occur. Waddington stressed the importance of cytoplasmic components and their influence on gene expression, yet transgenerational or maternal influences mediated by the transformation of cytoplasmic components from a pregnant female to daughter would not recognize epigenetics under the definition of Holliday because the patterns of expression of daughter are dependent and results from the transformation of components from the cytoplasm, such as prions, RNA and transcription factors, *etc.* These problems make the contradiction between Holliday's epigenetics and Waddington's epigenetics much more obvious (Meyers, 2011).

Structurally & Biochemical Advances in Mammalian DNA Methylation

DNA methylation expresses the most durable epigenetic alternation that plays a significant role in managing specific patterns of expression. Especially in mammals, DNA sequences get methylated on cytosine nucleotides which become methyl-cytosine; as a result, transcription is inhibited. These altered cytosine nucleotides are normally present next to guanine, as a result, two methylated cytosine nucleotides become diagonal to each other on opposite strands. Densely methylated parts of DNA having high densities of these so-called CpG islands are usually located next to starting point transcription. This change is offered by DNA-methyltransferases (DNMTs) that consist of DNMT3b and DNMT3a that methylate the cytosine on the *de novo* strand after replication. DNMT3a attach to chromatin and then activate histone H3K4 methylation alteration that inhibits the attachment of DNMT3L, and thereby represses DNA methylation in the active parts. Thus, DNA methylation is a way through which transcription can be regulated. CG methylation makes CG-rich regions inactivated, making CG rare in the genome and converting 5-methylcytosine to thymidine. The genome has many CG-rich regions in its promoter. These CG-rich regions are termed as CG islands. CG islands are not methylated and thus become targets for dysregulation in blood diseases and cancers. Unusual DNA methylation has been observed in tumor suppressor genes of myelodysplastic disease. In a surprisingly correlated process, proteins that attach to methylated DNA also form networks with proteins required for histones deacetylation. Therefore, when the DNA is in a methylated state, nearby histones are deacetylated, resulting in a compact, semi-permanently silent chromatin. Furthermore, demethylated DNA does not draw deacetylating enzymes to the histones, but it often attracts histone acetyltransferases, providing histones to remain acetylated and favoring transcription. Genome-wide analysis of DNA methylation patterns suggests some motifs with transcription factors in CG island. These motifs were found resistant to methylation due to these transcription factors imparting stability to these CG islands without even methylation. Despite several suggested mechanisms, demethylation mechanism during mammalian cell differentiation (Maclean, 1986).

In early embryogenesis, DNA mainly lacks methylation. Post-implantation, *de novo* methylation starts, mediated essentially by DNA (cytosine-5-)-methyltransferase-3beta (DNMT3B) and -3alpha (DNMT3A). When methylation alters CpG islands, methyl-binding proteins trigger a silencing cascade. The cascade involves deacetylation and methylation of H3K9 histone, allowing the linkage of Protein 1 (HP1) and chromatin to become tighter. Demethylation of denovo DNA occurs after DNA replication. DNMT1 quickly finds *de novo* DNA for its methylation. The result is the replication of methylation patterns and the maintenance of silencing. Epigenetic reprogramming deletes methylation patterns in adults during early embryogenesis (Conklin *et al.*, 2012).

Histone Modifications

Histone modification is one of the posttranslational modifications of histone proteins. There are different types of histone modifications, such as the addition of different groups (phosphate, methyl and acetyl) and some proteins (Small Ubiquitin-like Modifier (or SUMO), and ubiquitin) altering the chromatin structure with subsequent change in gene expression. These modifications affect various biological processes like chromosome packaging, transcriptional activation, and inactivation, DNA damage, and repair.

In almost all species, histone H3 is essentially acetylated at lysines 56, 23, 18, 14, and 9, methylated at arginine 2, and lysines 79, 36, 27, 9, and 4, and phosphorylated at ser28, ser10, Thr11, and Thr3. Histone H4 is essentially acetylated at lysines 16, 12, 8, and 5, methylated at lys20, arg3, and phosphorylated at ser1. Thus, quantitative detection of several histone alterations would offer useful information for a better understanding of the epigenetic control of cellular processes and the development of histone remodeling enzyme-targeted medicines.

Acetyl CoA provides an acetyl group for histone acetylation in the presence of the enzyme histone acetyltransferases. The acetylation process is closely associated with many other cellular processes like chromatin dynamics, cell cycle, RNA synthesis, differentiation, DNA repairing, silencing, apoptosis, and nuclear import. Histone acetyltransferases (HATs) not only catalyze but also regulate H3 and H4 acetylation. A total of 20 HATs have been recognized into five families, *i.e.*, MYST, GNAT1, P300/CBP, TAFII 250, SRC (Nuclear receptor coactivators). Histone deacetylases (HDACs) repression can enhance H3 acetylation, while HAT repression can reduce H3 acetylation.

Histone deacetylases catalyze the removal of an acetyl group from histone protein. Histone deacetylases have been classified into four classes, Class-I to class IV HDACs. Cancers cause disturbance in the acetylation process. Does H3 acetylate at its lysine amino residue help acetylation pattern studies, thus leading to a better understanding of the gene activation regulation process and development of HAT targeted drugs? Class-I HDACs include 8, 3, 2, and 1. Class-II HDACs consist of 10, 9, 7, 6, 5, and 4. Class-III enzymes, identified as sirtuins, require NAD+ cofactors and include SIRTs 1-7. The Class-IV enzyme, which includes only HDAC11, has characteristics of both Class-II and Class-I. HDAC repression represents important effects on cell cycle arrest, apoptosis, and differentiation in cancer cells. HDAC repressors are currently being developed as anticancer agents.

Epigenomics

Epigenomics is the study of the whole set of epigenetic changes on the genome of a cell, known as the Epigenome. The field is analogous to genomics and proteomics, which is the study of the genome and proteome of a cell. The Epigenome is similar to the brain that gives orders to the cells to becomemuscle cells, skin cells or eye cells, *etc.* It does not modify your DNA sequence; it just tells which genes will be expressed. The genome is similar to the muscle of the process that takes orders from the Epigenome.

Epigenomics shares many common features with other genomics fields in both methodologies and abstract purpose. Epigenomics seeks to characterize and identify epigenetic alterations on a global level, just like the study of the whole set of DNA in genomics or the whole set of proteins in a cell in proteomics. The reasoning behind making the epigenetic analysis on a global level is that inferences can be made regarding epigenetic alterations, which might be likely through analysis of particular loci. As in the other genomics fields, epigenomics depend on bioinformatics, which combines the disciplines of mathematics, biology, and computer science. However, whereas epigenetic alterations have been identified and studied for decades, these improvements in bioinformatics technology have allowed analyses on a global scale (Meyers, 2012);(Meyers, 2011).

CONCLUSION

This chapter presented a detailed description of epigenetics, the role of DNA, RNA, and chromatin. Moreover, details regarding regulation of epigenetic chromosomal inheritance, structural and biochemical advances in mammalian DNA methylation, modifications in histones, and epigenomics are discussed.

REFERENCES

Armstrong, L. (2013). *Epigenetics; Garland Science.*

Blazer, D. G., Hernandez, L. M. (2006). Genes, behavior, and the social environment. *Moving Beyond the Nature/nurture Debate.*

Conklin, K.F., Doerfler, W., Grafstron, R., Groudine, M., Hamilton, D., Jaenisch, R., Kruczek, I. (2012). *DNA methylation: biochemistry and biological significance.* Springer Science & Business Media.

Huang, S., Litt, M.D., Blakey, C.A. (2015). *Epigenetic gene expression and regulation.* Academic Press.

Maclean, N. (1986). DNA Methylation-biochemistry and biological significance. In: A., Razin, H., Cedar, AD., Riggs, (Eds.), *Bioessays* Springer-Verlag.(p. 392). New York:

Meyers, R.A. (2011). *Epigenetic Regulation and Epigenomics.* Weinheim: Wiley-VCH.

Meyers, R.A. (2012). *Epigenetic Regulation and Epigenomics.* John Wiley & Sons.

CHAPTER 6

Pan-Genomics

Abstract: This chapter proposes a brief description of the concept of pan-genomics; Computation approaches in pan-genomics and challenges; Applications of pan-genomics; Evolutionary pan-genomics and its perspectives; Pan-genomics for cancer prevention and its prospects; Pan-genomics of pathogens and its applications; Pan-genome analysis of microorganisms; Plant Pan-genomics.

Keywords: Microbial genome, Pan-genomics, Plant pan-genomics.

It was in 1995 when the first-ever microbial genome was sequenced to start the genomic era. The biggest achievement was the sequencing of the human genome in 2001. Immense genome sequencing has opened new horizons for microbial diversity. Genomics has helped discover the hidden gene repository in bacteria. The major aim of the pan-genome is to compare the genomes of distinct strains of the same species, and the concept can be extended to other taxa, such as genus.

Now, the accessibility of the wide number of genomes from various isolates of a similar pathogen has opened the likelihood of exploring various genomic attributes that are native to one or more species. One approach to examining these characteristics is through the pan-genomic approach. The first time the term pan-genome was coined by Tettelin and his colleagues in 2005. They utilized eight distinctive strains of *Streptococcus agalactiae*, a disease-causing bacteria isolated from humans. After this work, many other studies were conducted for pan-genome analysis of different microbes, including *Escherichia coli, Streptococcus pneumoniae, Bacillus cereus, Corynebacterium pseudotuberculosis, Corynebacter ium diphtheria* and many others. The thinking of pan-genomic studies leads to noteworthy bits of knowledge about bacterial evolution, host interaction, niche adaptation, and additional deductions in increasingly applied issues such as drug designing and vaccine development ("Computational pan-genomics: status, promises and challenges," 2018);(Sherman & Salzberg, 2020).

The term "pan-genome" can be defined as "global gene repertoire of a species". It reflects the full complement of genes present in a species, population, or given data set. Basically, the pan-genome comprises three main components. First is the core genome, second is the accessory genome, and then third is the strain-specific or species-specific genome. Core genome is the set of genes that are shared by all strains of a species and it plays a vital role in major cellular events: accessory or dispensable genome is comprised of genes that are present in some strains or absent in some isolates. Third, the strain-specific genome is the collection of unique genes that are specific to only one strain. The latter two are thought to be involved in niche adaptation.

To acquire the entire gene collection of given strain or species, it is important to decide what extra genes are included in recently sequenced genomes. Hence, the idea of open and closed pan-genome is very important. Some of the bacterial strains have a large gene pool, so they are said to have an open pan-genome, while some other species are limited to have a large gene pool, so they are said to have closed-pan-genome (Yuvaraj, Sridhar, Michael, & Sekar, 2017).

Open Pan-genome

In an open pan-genome, the number of genes increases as the number of sequenced strains increases. These strains have a maximum number of genes in their gene pool. Normally, these types of strains or species live in various conditions of blended microbial networks and have many ways to exchange their genetic material, so their gene pool expands continuously. *E. coli* has open pan-genome.

Closed Pan-genome

In a closed pan-genome, additional sequenced strains are not able to extend the number of genes in the gene pool because these types of strains or species tend to live in restricted and isolated niches and lack the mechanism for recombining and exchanging foreign genes between them. *Bacillus anthracis* has been known to contain a closed pan-genome.

Computational Approaches in Pan-genomics and Challenges

Nearly all biological science disciplines, including human genetics, microbiology, molecular biology, oncology, and virology, are facing the common challenge of analyzing the increasing number of sequenced genomes. In the coming few years, the number of sequenced human genomes will touch the value of hundreds of thousands or so, novel computational models and techniques are required to analyze such genomic data instead of any scaling up of pre-bioinformatics pipelines. Thus, there is a need for "Computational pan-genomics," a new subclass of computational biology research. Computational pan-genomics attracts a variety of applications beyond many disciplines, and thus, it connects with various other disciplines of bioinformatics like metagenomics, comparative genomics, and population genetics. These fields don't catch all parts of pan-genomics, yet they have built up their own algorithms and information structures to represent the genomic data and contribute to pan-genomics. By incorporating 'computational pan-genomics', we could build attention to regular challenges and create collaboration among the included fields. A dozen software tools are being used to profile core, accessory, and strain-specific genomes, identify SNPs, cluster orthologous genes, and build phylogenies. Each tool has its own advantages and limitations, providing room for improvement (Yuvaraj et al., 2017);(Pantoja, da Costa Pinheiro, Araujo, da Costa Silva, & Ramos, 2020). Some pan-genome analysis software tools that have different features are presented in Table **6.1**.

Table 6.1. Computational tools for pan-genome analysis.

Name	Main Characteristics	Reference
Panseq	Analysing core and dispensable genome and SNPs identification.	(Laing *et al.*, 2010)
PGAP	SNPs identification, clustering of genes, Function based analysis and constructing phylogeny of strains	(Zhao *et al.*, 2011)
PGAT	Clustering of genes, Function-based analysis and SNPs identification	(Brittnacher *et al.*, 2011)
ITEP	Clustering of genes, annotation, visualization, constructing phylogenies, and function-based analysis	(Benedict *et al.*, 2014)
PanCGHweb	Building phylogenies and clustering of genes	(Bayjanov *et al.*, 2010)
GET_HOMOLOGUES	Pan-genomic profile plotting, annotation, visualization and constructing phylogenies	(Contreras-Moreira and

(Table 6.1) cont.....

		Vinuesa, 2013)
CAMBer	Pan-genomic profile plotting and annotation	(Wozniak *et al.*, 2011)
PanCake	Analysing core and dispensable genome.	(Beißbarth *et al.*, 2013)
Harvest	SNPs identification, clustering of genes and visualization	(Treangen *et al.*, 2014)
PANNOTATOR	Analysing core and dispensable genome and annotation.	(Santos *et al.*, 2013)
PanGP	Pan-genomic profile plotting and visualization.	(Zhao *et al.*, 2014)

Applications of Pan-genomics

Pan-genome is the gateway towards many different applications, which can be explained by discussing reverse vaccinology, plant pan-genomics, and human pan-genome, and its contribution to human genetic diseases. Although the concept of pan-genomics is not limited to these domains, it has been believed that these three domains have been strongly impacted by pan-genomic methods (Liang, 2020).

It is believed that the conventional vaccine approaches are not proved to be successful for the pathogens that are only expressed in specific growth conditions or only expressed during infection. Their respective antigens are laborious to analyze. In this context, only an opportunity to deal with vaccine development from a particular aspect expresses all the protein antigens synchronously irrespective of abundance. For this, a new approach to reverse vaccinology endorses the computational approach to interpret complete genome sequencing data to find potential vaccine candidates *in-silico*. For the first time reverse vaccinology approach was practiced with *Neisseria meningitidis* whose pathogenic strain (MC58) was fully sequenced and evaluated to infer its convenient vaccine candidate by applying *in silico* approach. Cellular localization is a primary criterion to scrutinize an antigenic molecule of a microbial protein for vaccine development. For vaccination, the only surface-associated and secretary structures are considered to be immunological targets rather than cytosolic protein. The proteins that are considered to be the immunological targets presumed to have barred surface ensuring the availability of amino acid motifs that are liable for targeting the mature protein to the outer membrane (signal peptides), to the lipid bilayer (lipoproteins), to the integral membrane (hydrophobic transmembrane domains), or for recognition and interaction with host proteins or structures. Reverse vaccinology approach is quite successful for a variety of microorganisms and eukaryotic parasites, but having some constraints like the availability of a fully sequenced genome and *in vitro in vivo* testing models to interpret the antigen protective immune response (Pantoja *et al.*, 2017).

With the growing concept of pan-genome, variations between the species and in the same species have started to affect the genomic data of another genus. *Homo sapiens* is one of the eukaryotic species that have a huge number of genomic sequences. The foundation of human genomic studies was laid by the human genome project in 1990. Ensuing investigations revealed genomic variations and distinguished impacts on phenotypic contrasts in disease predisposition. According to the reference genome in databases, human genetic variations studies mainly focused on identifying and cataloging SNPs (single nucleotide polymorphism) and their association with the disease. Many other projects, including 1000 genomes project and U.K.'s 100,000 genomes project, revealed large public variation data leading to a conclusion that human genetics is not described better by reference sequence built previously. Now there is a time to reboot the human genome. Considerable exertion has been taken to refine the reference sequence, but still, it is important to pick up a complete comprehension of data that is individual specific. The first attempt to build a human pan-genome was carried out in 2009, and it includes two people and a human reference genome. Although compared to microbes, it was tiny; it was estimated that 40 Mb sequence that includes some protein-coding genes was absent from the human reference genome. Another team of scientists conducted a pan-genome study by randomly choosing two people, and they estimated that gene numbers varied from 73 to 87 genes because of copy number variation. Likewise, a scientist in Denmark attempts to build a human pan-genome by sequence data from 10 sets of Danish father-mother-child trios and identify hundreds of novel structure variants. These captivating discoveries and the constrained measure of data about the degree provoked scientists to construct the human pan-genome. Human pan-genome can be defined as a "nonredundant repertoire of all human DNA sequence present in the entire human population". Combining the core human genome with individual-specific sequences allows the formation of the human pan-genome, and it will be crucial for the comprehension of personal and medical genomics. Building a complete human pan-genome, huge sequencing data needs, sequencing of *de novo* assembly instead of microarray-based techniques to uncover common polymorphic sequences. With persistent advancement in sequencing innovation, sequencing is turning into a reasonable and moderate strategy for investigating an expansive number of human genomes, making it plausible to set up an increasingly exhaustive comprehension of the human genome, to make disclosures in medical genomics, and to grow new applications for personalized medicine (Barh, Soares, Tiwari, & Azevedo, 2020).

Evolutionary Pan-Genomics

The genome of humans was totally sequenced and amassed in 2001 as a kind of reference genome. The coming of cutting-edge next-generation sequencing strategies has been prepared for the resequencing of the whole population of the phylogenetic clade or a specific species in a limited duration of time with the least expense. Hence, after this innovative insurgency, there was a change in perspective in the idea of the genome from a solitary reference genome to a pan-genome. Pan-genome has three components, *i.e.*, core genome (available in all individuals), accessory genome (present in certain individuals), and unique genome, which is present in only one individual of the taxon. Genes of the core genome take an interest in the fundamental metabolic functions such as giving antimicrobial resistance in bacteria and housekeeping-like activities. Also, the core genome serves as a unit of the conserved genome in order to deduce evolutionary links among various strains of microorganisms (Anari, 2020).

The accessory genome is likely to undergo events of gene gain and gene loss and often face horizontal gene transfer that encourage variations in novel niches. The first since forever idea of the core genome was created by Tettelin *et al.* while working on the bacterial species, *Streptococcus agalactiae*. From that point forward, the exploration work on pan genomes was used for numerous prokaryotic and eukaryotic species. The core genome may likewise be characterized as a joined investigation of an assortment of genomic sequences deal as reference for specific species. With the inclusion of a new sample, pan-genomic studies can add to the size of the core and accessory pan-genome components. A fruitful report with respect to a core genome depends on the nature of reference assembly, annotation, and determination of proper individuals for study. In prokaryotes, the core component of the genome is responsible for vertical gene transfer and homologous recombination, while the variable part is responsible for horizontal gene exchange.

In prokaryotes, the core part of the genome is responsible for vertical transmission and homologous recombination, while the variable part is identified with horizontal gene exchange. In particular species, the core and accessory part of the genome may follow diverse evolutionary linkages. Largely, the core part gives a stable metabolic and genomic backing to the species, whereas the variable part is liable for great variety among people in a populace. Most of this variable part is confined to the adaptable genomic islands with an excess size of 10 kb. Accordingly, the attractive highlights of an ideal pan-genome are culmination (*i.e.*, incorporates every functional component), steadiness (*i.e.*, unique features), understandability (*i.e.*, incorporates genomic data of all individuals or species), and proficiency. A pan-genome can be classified as the core, accessory, and singleton parts (Fig. **6.1**). The developmental history of a species can be reproduced by utilizing the sequences of their genome. The developmental signs in the genome as the content of gene, common genetic markers, or SNPs might be helpful data during phylogenetic reconstruction for taxa. The unbiased hypothesis of molecular evolution has been so valuable in uncovering the activity of natural determination since it produces quantitative and falsifiable forecasts that can be tried by contrasting datasets. Given the multifaceted nature of powers molding the pangenome, it could be important to look outside hereditary genetics for likely methodologies: Pangenomes share numerous attributes with metacommunities, most remarkably the possibility that elements (species or species) are examined from a pool to shape discrete sets (genomes or communities) that share organic cohesiveness (pangenome or metacommunity). Metacommunity environment has created assemblage of hypotheses to see how metacommunities are collected and organized. This may assist analysts with disentangling the cycles shaping the structure of pangenomes (Tettelin & Medini, 2020).

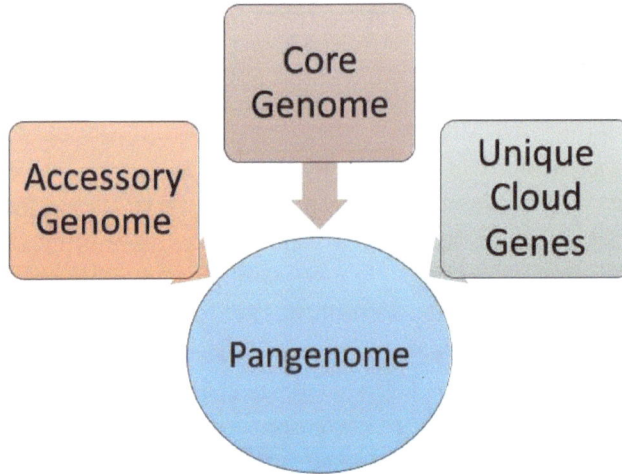

Fig. (6.1). Pictorial representation of components that come together to form pan-genome; at its center lies the core genome, which is present in all the strains or species, then comes the accessory genome consisting of genes which are present in one or more strains or species but not in all of them, finally, there are unique cloud genes which are strain-specific. Accessory and unique cloud genes are actually responsible for the inter and intraspecies variation.

Pan-genomics for Cancer Prevention and its Prospects

Cancer was realized as a carrier of genetic abnormalities in the early 1970s that can be explicit and periodic in nature. From there on, scientists have been enlightening the list of that mutation and genes that stimulate cancer and its progression for forty years. Innovative headways have incited this exploration by empowering genomic examination, with chromosome banding methods investigating the structure of the chromosome and also the cytogenetic procedures, positional cloning of genes of cancer, capillary sequencing that is the most mainstream strategy, comparative genomic hybridization, and the most recent being enormously used sequencing of the whole genome. With propelling innovation, the worldwide networks of cancer genomic societies set up The International Cancer Genome Consortium in 2008 for efficiently examining and reporting the mutations in somatic cells, including 25,000 samples covering the most widely recognized sorts of tumors. The fundamental purposes behind this worldwide consortium's development include an enormous extent of cancer examination, duplication of exertion if there are independent disease genomes. Various advancements utilized globally could represent a test in looking at datasets across studies, various sorts of tumors at different frequencies around the globe, and accessibility of information to the logical fraternity. Samples investigated by histopathologists of ICGC are cautiously prescreened and for clinicians to guarantee the nature of the sample and subsequently guarantee the exactness of the analysis. To meet the least inclusion and quality prerequisites, it is necessary to sequence both tumor and matched constitutional DNA. The information produced from ICGC is delivered quickly to the scientific society with the most extreme consideration to secure the moral and administrative viewpoints. The broad measure of information created from the projects of ICGC has empowered finding new disease genes related to cancer and their pathways, in explaining mutational cycles usable in human cancer, in an outline of the patterns of tumor heterogeneity (which is significantly seen in many tumors) and advancement of clonality, utilizing genomics for cancer prevention and clinical administration of malignant growth. The gigantic measure of information was investigated utilizing novel computational and statistical calculations explicitly intended to precisely distinguish genomic modifications and empower in clarifying new experiences in cancer. The gene expression data produced by TCGA has been made accessible for tumors and normal tissues of ten sorts of cancer that incorporate cancer of the breast, kidney renal papillary cell carcinoma, rectum adenocarcinoma, lung squamous cell carcinoma, and uterine corpus endometrioid carcinoma. This dataset is of most extreme significance in comprehending the natural mechanism of cancer and distinguishing proof of targets for new treatment (Table **6.2**). One of the significant uses of data on pan-cancer is to develop drugs by positioning medication targets that can be additionally exploited to advance targeted treatments for cancer.

Further investigation of the information is required for understanding interactions of gene-gene and functions of hereditary variations influencing pathways. An elective route for drug advancement could be pharmacological modulation of a significant pathway of cancer. In the coming years, when more knowledge is accessible from the information investigated, we may have data on extra tumor characterizations and different classifications of mutations like those found in the non-coding areas.

Table. 6.2 Web tools for the study of pan-cancer.

Name of Tool and Purpose	Weblink
IntOGen-mutations: The ID of drivers of cancer and its different types	http://www.intogen.org/mutations
CancerMiner: Recognizable proof of repeating miRNA-mRNA relationship over different types of cancer	http://cancrminer.org
Synapse: Sharing and refreshing information and results in collaboration with TCGA pan-cancer group	http://www.synapse.org/

(Table 6.2) cont.....

UCSC Cancer Genomics Browser: Gives an interactive investigation of genomic and clinical information	http://genome-cacner.ucsc.edu

Pan-genomics of Pathogens and their Applications

The pan-genome sequence information of a variety of organisms is readily available in public databases. Microbes attract more in constructing the pan-genomic map due to unique and small genomic size and the availability of closed pan-genomes of various microorganisms. Moreover, such bacterial species that are of clinical interest, their sufficient number of strains are sequenced to generate the assembly of the draft genome. Particularly, microbial pan genomics supports the studies based on comparative genomics because microorganisms are more prone to horizontal gene exchange. Microorganisms' potential towards horizontal gene transfer suggested that the genes present in one species' genome not all belong to the same phylogenetic tree. However, the microorganism's evolution depicts the phylogenetic network instead of a phylogenetic tree, and this phylogenetic network can be encoded in the Pan-genome framework. Pan-genome is promising in targeting the exact genetic variation of the respective trait by applying GWAS studies to microbes.

Reverse Vaccinology

For vaccine advancement, it is important to understand the pathogenesis of the disease. For this, biochemical, immunological, and microbiological methods are well understood. These conventional methods are proven helpful in recognizing many antigens, but sometimes it takes several years to characterize and consider the specific antigen for vaccine development. Moreover, these conventional approaches are not proven successful for pathogens only expressed in specific growth conditions or during infection. Their respective antigens are laborious to analyze. In this context, only an opportunity to deal with vaccine development from a particular aspect that expresses all the protein antigens synchronously irrespective of abundance. For this, a new approach reverses vaccinology that endorses the computational approach to interpret the data obtained from a pathogen's complete genomic sequence and infer potential vaccine candidates *in silico*.

For the first time reverse vaccinology approach was practiced to *Neisseria meningitides* whose pathogenic strain (MC58) was fully sequenced and evaluated to infer its convenient vaccine candidate by applying *in silico* approach. Cellular localization is a primary criterion to scrutinize an antigenic molecule of a microbial protein for vaccine development. For vaccination, the only surface-associated and secretary structures are immunological targets rather than cytosolic proteins.

The proteins that are considered to be the immunological targets presumed to be barred surface ensuring the availability of amino acid motifs that are liable for targeting the mature protein to the outer membrane (signal peptides), to the lipid bilayer (lipoproteins), to the integral membrane (hydrophobic transmembrane domains), or for recognition and interaction with host proteins or structures.

The reverse vaccinology approach is quite successful for the variety of microorganisms and the eukaryotic parasites, but having some constraints like the availability of a fully sequenced genome and *in vitro in vivo* testing models to interpret the antigen protective immune response (Luan & Thi, 2020).

Pan-Genome Analysis of Microorganisms

As day-by-day genome sequencing is becoming easy, scientists have evaluated different species' pan-genomes, including *rhamnosus*, *Corynebacterium diphtheria*, *Buchnera aphidicola*, *Pantoea ananatis* and *Corynebacterium pseudotuberculosis*. The properties of their pan-genome analysis with respect to pan-genome type, the number of genomes sequenced, and the total number of genes in pan-genome are given in Table **6.3**.

Table 6.3. Details of available pan-genomes for different species.

Species	Number of Genomes	Open/closed Pan-genome	Pan-Genome Size
Lactobacillus rhamnosus	13	Open	4893 genes
Corynebacterium diphtheriae	13	Open	4786 genes
Buchnera aphidicola	6	Closed	2597 genes
Pantoea ananatis	8	Open	5566 genes
Corynebacterium pseudotuberculosis	15	Open	2782 genes

As the concept of pan-genome is rising due to high genetic drift among the species, many strains of bacteria are under consideration for pan-genome analysis to characterize genes responsible for genetic diversity and antibiotic resistance (Medini, Donati, Tettelin, Masignani, & Rappuoli, 2005).

Plant Pan-Genomics

In 2000 first plant genome was sequenced, which allowed geneticists to carry out comparative studies using this genome as a reference genome. Most of these studies focused on single nucleotide polymorphisms for their easy identification. But with the introduction of pan-genomics, geneticists working with plant genomes thought that a single reference genome is insufficient to justify the diversity of genomes. This diversity exists because of the structural variations in genomes, such as copy number variations and presence/absence variations. Copy number variations or CNV's are defined as those sequences whose number of copies vary among individuals of the same species where CNV's incorporate insertions, deletions, and duplications. On the other hand, presence/absence variations or PAV's, on the other hand, are sequences that are present in some genomes and absent in others (Bayer, Golicz, Scheben, Batley, & Edwards, 2020).

The plant pan-genomics is based on studies of all structural variants, most important of which are CNV's and PAV's. Although there is less information on the effects of CNV's and PAV's on plant's phenotypes, some of their functions are known, including not limited to; nutrient uptake, metabolite production, determination of flowering initiation, and biotic stress response. Various studies have been carried out on determining the various level of the newly sequenced genome from the reference genome. In this regard, an early comparative study on divergent ecotypes of *Arabidopsis thaliana* revealed 3.4 Mb of genomic sequence to be extremely variable from the reference genome. Similarly, other studies on soybean, rice, maize, and others also revealed significant variation in their genomes. So, all this variation requires us to use a pan-genomic approach to storing and using a large amount of data as most plant genomes are quite complex and relatively large(Bennetzen, Flint-Garcia, Hirsch, & Tuberosa, 2018). For this purpose, various genomic projects for different crops and plants are in progress (Table **6.4**).

Using a pan-genomic approach for plant genomes will be fruitful in terms of applications related to plant breeding, as genomic hybridization has been extensively used to induce our desired traits in the offspring. Wild type genes, which can improve important properties of our crops ranging from pest resistance to drought and heat tolerance and nutrition levels, are even now incorporated in our commercial crops. So, the pan-genomic approach will help geneticists easily identify novel genes by comparing newly sequenced genomes with the available reference pan-genome (Facimoto, Balbo, Chideroli, & de Pádua Pereira, 2020).

Table 6.4. Rundown pan-genome studies of plants.

Name of Species	Approach Utilized	Status of Domestication	Ploidy	Accession Number	No. of Pan-genome genes	Refs
Brassica rapa	*De novo*	Crop	Diploid	3	41,858 genes	(Lin *et al.*)
G. soya (soybean)	*De novo*	Wild	Tetraploid	7	59,080 gene	(Li *et al.*)
O. sativa	*De novo*	Crop	Diploid	3	39,891 genes	(Schatz *et al*)
B. oleracea	Iterative assembly	Crop	Diploid	10	61,379 genes	(Golicz *et al.*)
B. distachyon	*De Novo*	Wild	Diploid	54	37,886 genes	(Gordon *et al.*)
Medicago truncatula	*De Novo*	Wild	Diploid	15	75,000 gene	(Zhou *et al.*)
B. napus	Iterative Assembly	Crop	Tetraploid	53	94,013 genes	(Hurgobin *et al.*)
O. sativa	Map-to-pan	Crop	Diploid	3010	48,098 genes	(Wang *et al.*)
Sesamum indicum	*De novo*	Under-utilized crop	Diploid	5	26,472 gene	(Yu *et al.*)
Juglans (walnut)	*De Novo*	Wild	Diploid	6	26,458 gene	(Trouern *et al.*)

CONCLUSION

This chapter presents a detailed description of the concept of pan-genomics, computational approaches in pan-genomics and challenges, and applications of pan-genomics. Moreover, details of evolutionary Pan-genomics, Pan-genomics for cancer prevention, and its prospects are also summarized. It also discusses Pan-genomics of pathogens and their applications, reverse vaccinology, Pan-genome analysis of microorganisms, including plant Pan-genomics.

REFERENCES

Anari, S.S. (2020). *Towards comparative pan-genomics*. Wageningen University.

Bayer, P.E., Golicz, A.A., Scheben, A., Batley, J., Edwards, D. (2020). Plant pan-genomes are the new reference. *Nat. Plants, 6*(8), 914-920.

Bayjanov, J.R., Siezen, R.J., van Hijum, S.A. (2010). PanCGHweb: a web tool for genotype calling in pangenome CGH data. *Bioinformatics, 26*(9), 1256-1257.

Beißbarth, T., Kollmar, M., Leha, A., Morgenstern, B., Schultz, A.K., Waack, S., Wingender, E., ßhauer, K.P., Meinicke, P., Bernt, M. and Wieseke, N., (2013). German Conference on Bioinformatics 2013.GCB'13, September 10–13, 2013, Göttingen, Germany. http://www.dagstuhl.de/dagpub/978-3-939897-59-0.

Benedict, M.N., Henriksen, J.R., Metcalf, W.W., Whitaker, R.J., Price, N.D. (2014). ITEP: an integrated toolkit for exploration of microbial pan-genomes. *BMC Genomics, 15*(1), 8.

Bennetzen, J., Flint-Garcia, S., Hirsch, C., Tuberosa, R. (2018). *The maize genome*. Springer.

Brittnacher, M.J., Fong, C., Hayden, H.S., Jacobs, M.A., Radey, M., Rohmer, L. (2011). PGAT: a multistrain analysis resource for microbial genomes. *Bioinformatics, 27*(17), 2429-2430.

Contreras-Moreira, B., Vinuesa, P. (2013). *Towards Comparative Pan-genomics*. Wageningen University.

Facimoto, C.T., Balbo, L., Chideroli, R.T., de Pádua Pereira, U. (2020). Pan-genomics of food pathogens and its applications. *Pan-genomics: Applications, Challenges, and Future Prospects* Elsevier.

Golicz, A.A., Bayer, P.E., Barker, G.C., Edger, P.P., Kim, H., Martinez, P.A., Chan, C.K., Severn-Ellis, A., McCombie, W.R., Parkin, I.A., Paterson, A.H., Pires, J.C., Sharpe, A.G., Tang, H., Teakle, G.R., Town, C.D., Batley, J., Edwards, D. (2016). The pangenome of an agronomically important crop plant Brassica oleracea. *Nat. Commun., 7*(1), 13390.

Gordon, S.P., Contreras-Moreira, B., Woods, D.P., Des Marais, D.L., Burgess, D., Shu, S., Stritt, C., Roulin, A.C., Schackwitz, W., Tyler, L., Martin, J., Lipzen, A., Dochy, N., Phillips, J., Barry, K., Geuten, K., Budak, H., Juenger, T.E., Amasino, R., Caicedo, A.L., Goodstein, D., Davidson, P., Mur, L.A.J., Figueroa, M., Freeling, M., Catalan, P., Vogel, J.P. (2017). Extensive gene content variation in the Brachypodium distachyon pan-genome correlates with population structure. *Nat. Commun., 8*(1), 2184.

Hurgobin, B., Golicz, A.A., Bayer, P.E., Chan, C.K.K., Tirnaz, S., Dolatabadian, A., Schiessl, S.V., Samans, B., Montenegro, J.D., Parkin, I.A.P., Pires, J.C., Chalhoub, B., King, G.J., Snowdon, R., Batley, J., Edwards, D. (2018). Homoeologous exchange is a major cause of gene presence/absence variation in the amphidiploid Brassica napus. *Plant Biotechnol. J., 16*(7), 1265-1274.

Laing, C., Buchanan, C., Taboada, E.N., Zhang, Y., Kropinski, A., Villegas, A., Thomas, J.E., Gannon, V.P. (2010). Pan-genome sequence analysis using Panseq: an online tool for the rapid analysis of core and accessory genomic regions. *BMC Bioinformatics, 11*(1), 461.

Li, Y.H., Zhou, G., Ma, J., Jiang, W., Jin, L.G., Zhang, Z., Guo, Y., Zhang, J., Sui, Y., Zheng, L., Zhang, S.S., Zuo, Q., Shi, X.H., Li, Y.F., Zhang, W.K., Hu, Y., Kong, G., Hong, H.L., Tan, B., Song, J., Liu, Z.X., Wang, Y., Ruan, H., Yeung, C.K., Liu, J., Wang, H., Zhang, L.J., Guan, R.X.,

Wang, K.J., Li, W.B., Chen, S.Y., Chang, R.Z., Jiang, Z., Jackson, S.A., Li, R., Qiu, L.J. (2014). *De novo* assembly of soybean wild relatives for pan-genome analysis of diversity and agronomic traits. *Nat. Biotechnol., 32*(10), 1045-1052.
http://dx.doi.org/10.1038/nbt.2979 PMID: 25218520

Liang, Q. (2020). *Methods for Comparative Genome Analysis With Applications to Pan-Genomics and Genome Annotation.* Riverside: University of California.

Lin, K., Zhang, N., Severing, E.I., Nijveen, H., Cheng, F., Visser, R.G., Wang, X., de Ridder, D., Bonnema, G. (2014). Beyond genomic variation--comparison and functional annotation of three Brassica rapa genomes: a turnip, a rapid cycling and a Chinese cabbage. *BMC Genomics, 15*(1), 250.

Luan, N.T., Thi, H.H.P. (2020). Pan-genomics of aquatic animal pathogens and its applications. *Pan-genomics: Applications, Challenges, and Future Prospects* Elsevier.

Medini, D., Donati, C., Tettelin, H., Masignani, V., Rappuoli, R. (2005). The microbial pan-genome. *Curr. Opin. Genet. Dev., 15*(6), 589-594.

Pantoja, Y., da Costa Pinheiro, K., Araujo, F., da Costa Silva, A.L., Ramos, R. (2020). Bioinformatics approaches applied in pan-genomics and their challenges. http://dx.doi.org/10.1016/B978-0-12-817076-2.00002-0

Pantoja, Y., Pinheiro, K., Veras, A., Araújo, F., Lopes de Sousa, A., Guimarães, L.C., Silva, A., Ramos, R.T.J. (2017). PanWeb: A web interface for pan-genomic analysis. *PLoS One, 12*(5)e0178154 http://dx.doi.org/10.1371/journal.pone.0178154 PMID: 28542514

Santos, A.R., Barbosa, E., Fiaux, K., Zurita-Turk, M., Chaitankar, V., Kamapantula, B., Abdelzaher, A., Ghosh, P., Tiwari, S., Barve, N., Jain, N., Barh, D., Silva, A., Miyoshi, A., Azevedo, V. (2013). PANNOTATOR: an automated tool for annotation of pan-genomes. *Genet. Mol. Res., 12*(3), 2982-2989.

Schatz, M.C., Maron, L.G., Stein, J.C., Hernandez Wences, A., Gurtowski, J., Biggers, E., Lee, H., Kramer, M., Antoniou, E., Ghiban, E., Wright, M.H., Chia, J.M., Ware, D., McCouch, S.R., McCombie, W.R. (2014). Whole genome *de novo* assemblies of three divergent strains of rice, Oryza sativa, document novel gene space of aus and indica. *Genome Biol., 15*(11), 506. PMID: 25468217

Sherman, R.M., Salzberg, S.L. (2020). Pan-genomics in the human genome era. *Nat. Rev. Genet., 21*(4), 243-254.

Tettelin, H., Medini, D. (2020). The pangenome: Diversity, dynamics and evolution of genomes. *Publisher Springer Nature.* https://www.springernature.com/gp/products/books

Treangen, T.J., Ondov, B.D., Koren, S., Phillippy, A.M. (2014). The Harvest suite for rapid core-genome alignment and visualization of thousands of intraspecific microbial genomes. *Genome Biol., 15*(11), 524.

Tobias Marschall, Manja Marz, Thomas Abeel, Louis Dijkstra, Bas E. Dutilh, Ali Ghaffaari, Paul Kersey, Wigard P. Kloosterman, Veli Makinen, Adam M. Novak, Benedict € Paten, David Porubsky, Eric Rivals, Can Alkan, Jasmijn A. Baaijens, Paul I. W. De Bakker, Valentina Boeva, Raoul J. P. Bonnal, Francesca Chiaromonte, Rayan Chikhi, Francesca D. Ciccarelli, Robin Cijvat, Erwin Datema, Cornelia M. Van Duijn, Evan E. Eichler, Corinna Ernst, Eleazar Eskin, Erik Garrison, Mohammed El-Kebir, Gunnar W. Klau, Jan O. Korbel, Eric-Wubbo Lameijer, Benjamin Langmead, Marcel Martin, Paul Medvedev, John C. Mu, Pieter Neerincx, Klaasjan Ouwens, Pierre Peterlongo, Nadia Pisanti, Sven Rahmann, Ben Raphael, Knut Reinert, Dick de Ridder, Jeroen de Ridder, Matthias Schlesner, Ole Schulz-Trieglaff, Ashley D. Sanders, Siavash Sheikhizadeh, Carl Shneider, Sandra Smit, Daniel Valenzuela, Jiayin Wang, Lodewyk Wessels, Ying Zhang, Victor Guryev, Fabio Vandin, Kai Ye and Alexander Scho"nhuth. (2018) Computational pan-genomics: status, promises and challenges. *Briefings in Bioinformatics*, 19(1), 118-135.

Trouern-Trend, A.J., Falk, T., Zaman, S., Caballero, M., Neale, D.B., Langley, C.H., Dandekar, A.M., Stevens, K.A., Wegrzyn, J.L. (2020). Comparative genomics of six Juglans species reveals disease-associated gene family contractions. *Plant J., 102*(2), 410-423.

Wang, W., Mauleon, R., Hu, Z., Chebotarov, D., Tai, S., Wu, Z., Li, M., Zheng, T., Fuentes, R.R., Zhang, F., Mansueto, L., Copetti, D., Sanciangco, M., Palis, K.C., Xu, J., Sun, C., Fu, B., Zhang, H., Gao, Y., Zhao, X., Shen, F., Cui, X., Yu, H., Li, Z., Chen, M., Detras, J., Zhou, Y., Zhang, X., Zhao, Y., Kudrna, D., Wang, C., Li, R., Jia, B., Lu, J., He, X., Dong, Z., Xu, J., Li, Y., Wang, M., Shi, J., Li, J., Zhang, D., Lee, S., Hu, W., Poliakov, A., Dubchak, I., Ulat, V.J., Borja, F.N., Mendoza, J.R., Ali, J., Li, J., Gao, Q., Niu, Y., Yue, Z., Naredo, M.E.B., Talag, J., Wang, X., Li, J., Fang, X., Yin, Y., Glaszmann, J.C., Zhang, J., Li, J., Hamilton, R.S., Wing, R.A., Ruan, J., Zhang, G., Wei, C., Alexandrov, N., McNally, K.L., Li, Z., Leung, H. (2018). Genomic variation in 3,010 diverse accessions of Asian cultivated rice. *Nature, 557*(7703), 43-49.

Wozniak, M., Wong, L., Tiuryn, J. (2011). CAMBer: An Approach to Support Comparative Analysis of Multiple Bacterial Strains.

Yu, J., Golicz, A.A., Lu, K., Dossa, K., Zhang, Y., Chen, J., Wang, L., You, J., Fan, D., Edwards, D., Zhang, X. (2019). Insight into the evolution and functional characteristics of the pan-genome assembly from sesame landraces and modern cultivars. *Plant Biotechnol. J., 17*(5), 881-892.

Yuvaraj, I., Sridhar, J., Michael, D., Sekar, K. (2017). PanGeT: Pan-genomics tool. *Gene, 600*, 77-84.

Zhao, Y., Jia, X., Yang, J., Ling, Y., Zhang, Z., Yu, J., Wu, J., Xiao, J. (2014). PanGP: a tool for quickly analyzing bacterial pan-genome profile. *Bioinformatics, 30*(9), 1297-1299. http://dx.doi.org/10.1093/bioinformatics/btu017 PMID: 24420766

Zhao, Y., Wu, J., Yang, J., Sun, S., Xiao, J., Yu, J. (2012). PGAP: pan-genomes analysis pipeline. *Bioinformatics, 28*(3), 416-418. http://dx.doi.org/10.1093/bioinformatics/btr655 PMID: 22130594

Zhou, P., Silverstein, K.A., Ramaraj, T., Guhlin, J., Denny, R., Liu, J., Farmer, A.D., Steele, K.P., Stupar, R.M., Miller, J.R., Tiffin, P., Mudge, J., Young, N.D. (2017). Exploring structural variation and gene family architecture with *De Novo* assemblies of 15 Medicago genomes. *BMC Genomics, 18*(1), 261.

CHAPTER 7

Digital Genomic Era

Abstract: This chapter proposes a brief description of the Role of Bioinformatics in the Post-Genomic Era; Digital Sequence Repositories; From the World of 1D to the era of 3D (Protein Structures); Functional Genomics; Sequence Alignment, Phylogeny, and concept of gene evolution; Sequence and structural alignment; Genomics of personalized medicine; CADD: Computer-Aided Drug Designing; Big Data Analysis; Responsible Genomic data sharing.

Keywords: Alignment, Big data, CADD, Digital Sequence Repositories, Functional Genomics, Protein Structures.

ROLE OF BIOINFORMATICS IN POST-GENOMIC ERA

Bioinformatics has had a constitutive role in research and biomedical sciences for the past few decades. Now, bioinformatics has an integral role in analyzing genomics, transcriptomics, and proteomics data created by high-throughput technologies. For analyzing single genes and proteins, sequence-based methods have been expanded and elaborated. Sequence-based methods have also been developed for large sets of genes or proteins concurrently. As we have the whole genome of each organism at hand, bioinformatics plays a major role in detecting the systematic functional behaviour of cells and the organism. Post-genomic era started from the completion of the human genome project (Torshin, 2006). The role of bioinformatics in post-genomics era in many areas is described as:

- Completion of the human genome project
- Automatic annotation of genomes
- Comparative genome analysis
- Transcriptome analysis and discovery of genes.
- Analysis of gene expression and proteins motifs, domain findings
- Gene regulation network and System biology

Completion of Human Genome Project (HGP)

The first draft of HGP was announced in 2000. It was officially published and

Maryam Javed, Asif Nadeem & Faiz-ul Hassan

completed in 2003. Except a small number of genomes, good quality sequences have been obtained and deposited in different databases. NCBI, EBI, and other databases answer our questions, such as where our required gene is located, how many genes can be spliced, and the location of genetic markers on the genome. Genes in the vicinity can be studied which have a part in phenotypes. This information is available on the NCBI homology map for the identification of proteins having domains and motifs. Ensembl provides data mining systems which are called Ensmart. With the help of this novelty, Gene expression, SNPs and other complex queries can be built. These browsers are available not only for the human genome but for rats, mouse or others for which complete genome is available.

Automatic Annotation of Genomes

Ensembl genome database project provides a server for automatic annotation of human genome sequences. It anticipated genes based on:

- Protein evidences.
- Provide huge information about proteins sequences, structure & function.

For microbes, an automatic annotation system is known as a comprehensive microbial resource (TIGR/CMR *URL)* developed. Annotation information is provided by this system that is regularly updated to include new information added in the relevant database.

Comparative Genome Analysis

A comparison of genome sequences of a relevantly isolated species can be manipulated in detecting exons and other regulatory elements. This analysis can be done based on homology with the closely related species. For example, in detecting human genes, the genome sequences of the mouse are very useful. With the advent of technology, DNA sequencing price is becoming cost-effective and new sequences will be available for sequencing in the future. Proteins are conserved in some species are important for the biology of species and classification of species base on the profile of conserved sequences. Some associated tools are important in comparative analysis. **COG** (cluster of orthologous groups of proteins) provides the information mentioned above for microorganisms, and **KOG** (Eukaryotic orthologous group of proteins) compares the proteome of *S. cerevisiae, human, Arabidopsis,* and *Drosophila.* It is an important tool for domain structure detection in unique proteins. Almost 1.3 million sequences of non-redundant protein databases are managed by NCBI and over 19000 proteins by PDB.

Transcriptome and Gene Expression Analysis

The cDNA gives the experimental proof for the transcription. The cDNA-based approaches are important, and Transcriptome plays a vital role in gene expression, gene regulation, genomics, and proteomics. According to Ensembl, 28000 genes, each in the mouse and human genome, is identified when cDNA clones are collected and annotated.

DNA microarray technologies have become a famous tool for gene expression profiles, and the statistical methods linked with them are well explained. Samples are classified based on expression profile and cluster of genes identification that characterizes the sample. With bioinformatics tools, deregulated genes can be traced. The KEGG pathway database accepts the list of genes symbol and constructs the pathway of genes involved.

Gene Microarray Pathway Profiler (GenMAPP) gives us a statistical analysis and also visual presentation. **EXPANDER** is used to find the set of genes in upstream regions that share common promoter elements. Short motifs that can be found in numerous proteins play a role in the structure and function of proteins. Many machine learning approaches such as hidden Markov Models and other databases such as Pfam, EBI merge domain and motifs.

Computational methods have become more constitutional to recent biological research, which involves biologists and computer scientists. It is a concerted interdisciplinary cooperation between the two scientific communities, and we can say that bioinformatics is the need of the post-genomic era.

DIGITAL SEQUENCE REPOSITORIES

DNA Databases

DNA databases are used for analyzing genetic diseases and genetic fingerprinting. These databases can play a major role in the identification of missing individuals across time.

- DNA data bank of Japan (DDJB)
- European bioinformatics institute (EMBL)
- GeneBank (National Centre for biotechnology information NCBI)
- UniGene (database of the Transcriptome)

These databases are known as primary databases for the nucleotide sequence data. The DDJB is the member of INSD to assemble and give nucleotide sequence data with NCBI(USA) and ENA/EBI (Europe). The main purpose of DDJB is to upgrade international nucleotide sequence database collaboration (INSD) (Ziemann, Kaspi, & El-Osta, 2019).

Protein Sequence Databases

The databases of proteins are the accumulation of sequences from various sources and translations from annotated regions coded in Refseq, GenBank, and data from Swissprot, PDB, and PRF. These are the following protein data repositories.

- InterPro (Classify proteins into families and anticipate sites& domains)
- Swiss-Prot (To give authentic proteins sequences that are linked with higher annotation
- **PROSITE** (Proteins families &domains database)
- SUPERFAMILY (database of structural &functional annotation for all the genome & proteins)
- Pfam (database of alignment of protein families)

Proteins Structure Databases

Protein structure databases consist of the protein data bank in Japan, protein DataBank in Europe, and Research Collaboratory for Structural Bioinformatics. These databases aim to assemble and annotate the proteins structures. We can find the 3-D structure of proteins, which can be used to find an evolutionary relationship that cannot be found only by sequences comparison. Here are few examples of protein structure databases.

- The PDBe (database for macromolecular structures MSD. The analysis of PDBe allow the inspection of structural data parameters over the PDB rapidly) (Fragai, Luchinat, & Parigi, 2006).

- ModBase (database of 3-D proteins models that have been calculated by comparative modelling).
- Protein common interface database (ProtCID database for protein-protein interface in a crystal structure of the homologous proteins).

Comparative Genomics Databases

In comparative genomics, various genomic features such as DNA sequences, RNA, Gene order, and regulatory sequences of different organisms are compared. The database of comparative genomics is:

Cluster of Orthologous Groups (COGs)

It is based on the phylogenetic classification of proteins that are encoded in a complete genome. A cluster of orthologous groups can also construct the references sequences of bacterial and archaeal genomes.

Genomics Databases

Genomics databases are online repositories of genomic variants reported for locus-specific (single) or general genes (more genes) or precisely for one population. There are few databases of genomics:

CyanoBase

This is the genomic database of cyanobacteria which is used as a model organism for photosynthesis.

EcoGene Database

This database gives the number of proteins and genes sequences that have been derived from *E. coli* K-12. This database can do genetic and physical map compilation with the alliance of Coli Genetic stock Centre.

Mendel Database

Mendel database consists of names for plant-wide families of the sequenced plant genome.

RNA Databases

These are RNA Databases:

Pseudo Base

This database consists of sequenced, structural, and functional data that is associated with RNA pseudoknots (structural motifs present in the RNA).

Ribosomal Database Project

Provide the aligned & annotated ribosomal RNA genes (rRNA) sequenced data in addition to tools that allow the researchers to examine their own ribosomal RNA genes sequences in the ribosomal Database project (RDP) framework.

FROM THE WORLD OF 1D TO THE ERA OF 3D (PROTEIN STRUCTURES)

The molecular structure of the macromolecules has a champion role in modern-day research and technologies. Watson and Cricks predicted the DNA structure which contributed to the Molecular Biology field. Nowaday structure biology field gives huge knowledge about DNA, RNA, Protein, Ligand, and binding sites of the different molecules with the aid of modern-day technologies, like NMR, X-rays crystallography, and electron microscopy. Nowadays, the Worldwide Protein Data Bank contains a lot of protein structures determined through different experiments. The protein structures were generated through molecular modelling and simulation techniques. The benefits and drawbacks of 1D, 2D, and 3D visualization of the macromolecules were elaborated one by one.

1D TECHNIQUES

The visualization of the sequences of different amino acid residues with one latter abbreviation encoded the single nucleotide polymorphism, or secondary structure sequences, fell under the category of 1D technique (Fig.**7.1**). The sequences of the amino acid can be represented with different colours as well as stack. The multiple sequences can be aligned with different colours. The feature can be highlighted using different glyphs. The position of atoms in the 1D technique can not be represented in the spatial structure of the protein. Therefore, 1D technique can not complete the tasks related to spatial protein structure.

2D TECHNIQUES

The sequences of the amino acid or components in a molecule can be placed into two coordinates of the plane. The visualization of the molecule's structures can be represented in distance matrix and 2D projection. In this technique, 2D representation of the macromolecules lacks the complete information of the spatial conformation. The macromolecules can be visualized and analysed with the three most common 2D methods. The 2D drawing of the macromolecules through Lewis representation provides information about the bonding between the atoms. The second common method of 2D representation can be done through a distance matrix in which each amino acid residue or atom in a molecule is used to construct a distance matrix. The third common method of 2D representation is a topology diagram in which the secondary structure of the protein can be schematically represented. The secondary structure feature can be highlighted using different glyphs. Topology diagram is basically the connection between 1D and 3D structure of the molecules. Topology model is unable to show the 3D structure of the protein.

3D TECHNIQUES

The sequences of amino acid and compound components can be visualized in a 3D model of the macromolecules containing all the information necessary for spatial conformation of the structure. The 3D models of the macromolecules can be generated through different approaches. Ribbon diagram represented the 3D models of the different proteins. Space-filling and wireframe were also suitable approaches to 3D modelling. 3D balls and sticks are also falling under the umbrella of 3D modelling.

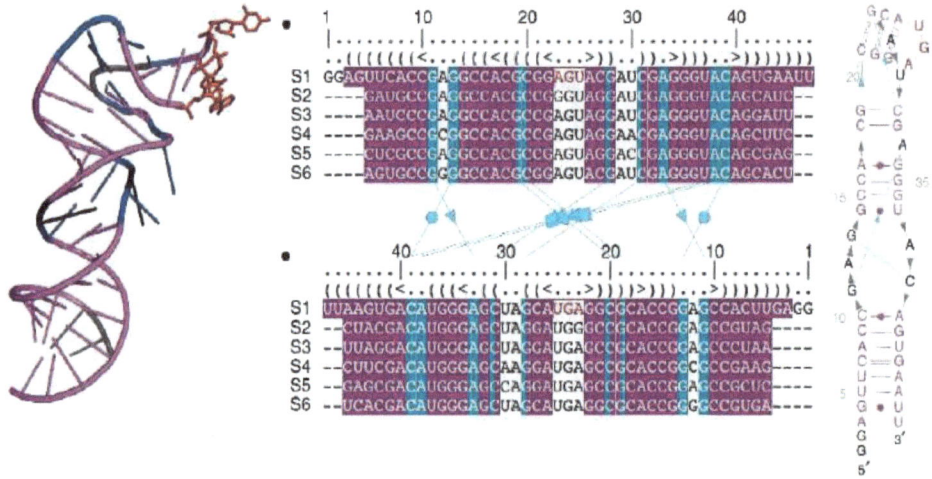

Fig. (**7.1**). The macromolecules structure of 1D (centralized), 2D (Right), and 3D (Left) visualization.

Functional Genomics

The term defines the functions of genes and their interaction. It uses data from sequencing projects of the genome and ribonucleic acid sequencing (RNA). It emphasizes making messenger RNA from deoxyribonucleic acid (DNA) strand and from this mRNA to protein and genes expression regulation and interaction between Protein-protein. An organism's genes and proteins are studied in functional genomics (FG), which also includes the study of the properties of gene products at biochemical and cellular levels. It also comprises genetic variation over time and disruptions related to functions as mutations. Producing knowledge at the proteomic or genomic level is also the main objective of FG. The data from the (FG) will provide complete knowledge about the comparison between genome and the single gene that how they specify their function.

Firstly, it is important to define a function to understand (FG). There are two ways to define it. One is the selected effect (SE), and the other is a causal role (CR). The function that refers to selecting a particular trait, i.e., DNA, RNA, or protein, to perform a function is called SE. A trait that is necessary and sufficient for a particular function is defined as CR. The definition of a function that is usually tested under the umbrella of FG is CR.

Multiplex techniques have been used in functional genomics to estimate the products of the gene in a sample (Pevsner, 2015). Some of the techniques that are used in functional genomics are given below.

At the DNA Level

DNA/Protein Interactions

mRNA have coded information for the synthesis of protein from DNA that shows the main effect on the regulation of gene expression. DNA sequences are essential in understanding the regulations of gene expression. To identify interaction sites for DNA and protein, the following techniques have been formed, as Chip-sequencing and CUT&RUN.

DNA Accessibility Assays

Assay for Transposase-Accessible Chromatin using sequencing, DNase-Seq, and Formaldehyde-Assisted Isolation of Regulatory Elements (FAIRE-Seq) techniques are used to detect sites of the genome that are accessible.

At Ribonucleic Acid (RNA) Level

Microarrays

Estimation of mRNA amount in a given sample relates to a particular gene.

Serial Analysis of Gene Expression (SAGE)

An alternate method for analysis for which RNA sequencing is the basis despite hybridization that includes microarrays. It is based on ten to seventeen base pairs sequencing that is distinctive to every single gene.

RNA Sequencing

It is the best method for studying the copying of mRNA from DNA (transcription) and gene expression.

Massively Parallel Reporter Assays:

Cis-regulatory activity of DNA sequences are checked by using this method.

At Protein level

Yeast Two-hybrid Screening

This method is used to know about the Protein-DNA interactions and protein-protein physical interactions.

Affinity Purification and Mass Spectrometry

This method detects those proteins that form complexes by interacting with one another.

Loss-of-function Techniques

Mutagenesis

With the help of deletion and disruption of function, the genes are deleted in this method. The resultant organism is selected for traits that offer hints to the purpose of the gene disruption.

RNA Interference

In this twenty base-pair (bp) double-stranded (ds) RNA characteristically carried by transfection of synthetic 20-mer short-interfering RNA molecules (siRNAs) or via virally encrypted short-hairpin RNAs to knock down gene expression using RNAi screens. This is regularly done in cell culture-based assays or experimental animals.

CRISPR Screen

The genes are deleted in a multiplexed way in cell lines by using CRISPR-Cas9.

Functional Annotations for Genes

In this genome, annotation and Rosetta stone approach are used.

Consortium Projects Focused on Functional Genomics

In this the Encyclopedia of DNA elements) and the Genotype-Tissue Expression (GTEx) are used.

Sequence Alignment, Phylogeny, and Concept of Gene Evolution

A sequence alignment is an approach to organizing the sequences of deoxyribonucleic acid (DNA), ribonucleic acid (RNA), or protein to find areas of resemblance resulting from structural, evolutionary, and functional interactions between the sequences. Adjusted groupings of nucleotide are commonly characterized as rows in the matrix. Gaps are embedded between the remains, so comparative characters are adjusted in progressive columns. Non-biological sequences are also done by sequence alignments. If a sequence is mismatched, this interpretation is point mutation: a single nucleotide base is changed, inserted, or deleted from a DNA or RNA sequence. In the case of proteins, the degree of resemblance among amino acids (AA) in a specific position can be understood as a conserved region. Even though DNA and RNA nucleotide bases are more similar than (AA), the conservation of base sets can show a comparative structural job.

Extremely short or fundamentally, the same sequence can be adjusted by hand. In any case, most intriguing issues require the arrangement of long or very vast groupings that can't be adjusted exclusively by human work. Human information is applied in developing calculations to create excellent succession arrangements and periodically changing the end-product to reflect designs that are hard to speak algorithmically (particularly on account of nucleotide groupings). Computational ways to deal with the arrangement, for the most part, fall into two classifications. One is global alignment (GA) and the other is local alignment (LA). By contrast, local alignments classify areas of similarity within long sequences that are extensively different. LA is consistently best; however, they can be harder to compute given the extra challenge of recognizing the similarity regions. An assortment of computational calculations has been applied to the problem of sequence alignment. These incorporate moderate yet officially right techniques like powerful programming. These additionally incorporate productive, heuristic calculations or probabilistic techniques intended for enormous database search.

The relationships between unlike groups of organisms and their evolutionary development are under the study of Phylogeny. Phylogeny tries to follow the developmental history of all life on the planet. It depends on the phylogenetic theory that all living beings share a joint heritage. The connections among organisms are portrayed in what is known as a phylogenetic tree. Relations are dictated by shared qualities, as shown through the correlation of hereditary and anatomical similarities. In phylogeny, DNA and protein structure investigation is utilized to decide hereditary relations among various organisms. For instance, the study of cytochrome C, a protein in cell mitochondria that capacities in the electron transport system and energy creation, is utilized to decide levels of relationship among organism's dependent on resemblances of amino acids (AA) sequence (seq.) in cytochrome C. Likenesses in qualities of biochemical structures, for example, DNA and proteins, are used to build up a phylogenetic tree dependent on acquired shared attributes.

A phylogenetic tree, or cladogram, is a schematic picture utilized as a visual representation of proposed developmental connections among taxa. Phylogenetic trees are diagrammed dependent on presumptions of cladistics or phylogenetic systematics. Cladistics is an arrangement framework that classifies living beings on shared attributes, or synapomorphies, as controlled by hereditary, anatomical, and molecular investigation. Hereditary researchers have accordingly changed this idea, for example, George C. Williams' gene-centric view of evolution. He proposed an evolutionary idea of the gene as a unit of natural selection with the definition: "that which segregates and recombines with considerable recurrence." In this view, the gene interprets as a unit, and the hereditary gene acquires as a unit-related thought, stressing genes' centrality in evolution.

For example, the investigation of taste receptor gene evolution has given key bits of knowledge into the dynamic course of variation of various species to their adaptation. Invertebrates, for example, bugs, have aversive reactions to unpleasant tasting or hindrances synthetic compounds. A Drosophila Gr66a taste receptor answerable for aversive reaction to caffeine, a compound that preferences unpleasant to people, is random to vertebrate T2R genes. This recommends that receptors for the identification of aversive taste evolved autonomously in vertebrate and invertebrate species. Fish and chickens have ten or less than ten T2R genes and no pseudogenes. Warm-blooded animals and frogs have much larger quantities of the T2R gene twenty-one to sixty-four and a great extent of T2R pseudogenes. A study of relatedness of the T2R gene in various species recommends a perplexing evolution of this gene family (Rosenberg, 2009).

SEQUENCE AND STRUCTURAL ALIGNMENT

Sequence alignment is a method of orchestrating DNA, RNA, or protein sequences to distinguish similar regions that might result from the structural, functional, and evolutionary linkage between the sequences. Sequences of aligned nucleotides or residues of amino acids are commonly spoken to as rows inside the matrix. Gaps are embedded between the subunits to ensure that characters that are identical in progressive columns. Sequence alignments are additionally utilized for nonorganic sequences, like computing the cost of the distance between strings in a characteristic language or monetary data. Sequence alignments are valuable in bioinformatics for distinguishing sequences' likeness, creating phylogenetic trees, and creating homology models of the protein structures. However, the biological significance of sequence alignment isn't, in every case, clear. Alignments are regularly expected to mirror developmental change between sequences descended from a common precursor; in any case, it is officially conceivable that convergent evolution can happen to deliver clear likeness between developmentally disconnected proteins but perform comparative functions and have comparable structures. As of now, comparative modeling or homology is precise and hence the most broadly utilized approach for predicting protein structure. Homology modeling is an empirically based observation that such evolutionarily related proteins will generally have the same three-dimensional structures. In addition, structural features of protein frequently remain preserved long after the signal of a sequence is lost to deletions, mutations, and insertions. Like this, a three-dimensional structure is viewed as the most powerfully preserved property of proteins that are homologous, positively more preserved than the molecular function or the sequence. Even though there are some persuading exemptions for this standard, it quite holds for the outright majority of the cases.

Homology modeling is utilized to construct a three-dimensional model of a protein-based on the sequence alignment of its amino acids with a known structure of a related protein. Each homology modeling strategy comprises four fundamental steps. Firstly, recognition of proteins that are related to each other and have structures that are experimentally determined and hence can be utilized as a structural format for modeling. Secondly, residues of the corresponding mapping between the structure of the template and the sequence of the target, process alluded to as sequence-structure alignment. Thirdly, creating a three-dimensional structure of a targeted protein based on the sequence-structure alignment and assessing the accuracy of the subsequent model. To obtain a satisfactory assessed quality, the entire process might be repeated until the model is no longer improved.

A complete rundown of accessible programming software arranged by algorithm and type of alignment is accessible at sequence alignment programming software. Yet, regular programming software utilized for general alignment of sequences incorporate ClustalW2 and T-coffee, and BLAST, FASTA3x for the searching database. Marketed available tools, such as DNASTAR Lasergene, Geneious (Bioinformatics Software for Sequence Data Analysis), and Pattern Hunter, are also accessible. Devices annotated as performing alignment of sequences are recorded in the bio-tools vault. Programming software and Alignment algorithms can be straightforwardly contrasted with each other utilizing a normalized set of benchmark references known as BAliBASE. The data set comprises structural alignments, which can be viewed as a standard against which methods are compared based on the sequence. The overall performance of numerous basic alignment strategies on often experienced alignment issues has been organized, and chosen results are distributed online at BAliBASE. A complete rundown of BAliBASE scores for various distinct alignment tools can be figured inside the protein workbench STRAP (Shanker, 2018).

GENOMICS OF PERSONALIZED MEDICINE

Personalized medicine is referred to medical treatment of the individual characteristics using genetic information of the patient, not only to improve diagnostic and disease treatment but also a potential role in early detection and treatment of the disease. The advancement of emerging technologies like DNA sequencing, proteomics, wireless health monitoring devices, and imaging protocols have a different response among individuals regarding mechanisms and effects during disease processes. Different individuals respond to different effects during disease treatment due to unique molecular, environmental exposure, physiological, and behaviour levels. Personalized medicine is the only alternative way to overcome disease-related problems by using individual genetic profiles. The individual can get benefits from the personalized medicine by their genomic information. Personalized medicine has a tremendous role in health care of mankind. It is designed according to the individual genetic, proteomics, disease diagnostic and treatment. By using genetic information, individuals can prevent themselves from the side effects. Genomic data of the individual helps the health scientist to choose appropriate therapy for disease treatment.

Personalized medicine leads to the development of pharmacogenomics. In pharmacogenomics, the effects of genes against particular drugs can be studied. This branch helps scientists study the effect of the medicine on responder and non-responder, drug dosages, and adverse effects on the individual. Genes and proteins were used as a biomarker in personalized medicine to dragonize the disease. Moreover, the genomics of the individual also predicts the individual variability to a specific drug. The individual response helps the pharmacist to prescribe a suitable drug against a specific disease. The genomics of the individual has a landmark role in drug designing and medical treatment.

The personalized medicine patients can be categorized based on genetic polymorphism of cytochrome p450 and enzyme polymorphism. The genomic information of patients helps to predict gene-drug interaction. The optimized drug efficacy and adverse effects can be studied through gene-drug interaction. The gene associated with metabolism and the immune system can be studied through gene-drug interaction. Gene-based drug targeting can also be studied through individual genomics. The molecular mechanism for a specific disease treatment was also studied through a gene-based drug targeting approach. The prediction and disease diagnostic were accurately studied through genomics information. The asthmatic patients were treated with the inhalation of the corticosteroids and β2-adrenergic. The patient of asthma showed less responsiveness and resistance to these treatments. The genomic of the asthma patient helps the scientist to study β2-adrenergic receptors for better treatment.

The genomic of personalized medicine have a tremendous benefit for drug designing rather than trial and error method. The life-threatening adverse effects can be minimized through personalized medicine. The cost of the clinical trial of a drug on animal models and human trials can be reduced through the genomics of individual personalized medicine. The failure of the drugs can be easily identified through genomic information. Moreover, the favorable response of the drugs can also study through genomics. The efficacy of the drugs can be improved with the aid of the genomics of personalized medicine (Ginsburg & Willard, 2009).

CADD: COMPUTER-AIDED DRUG DESIGNING

Drug development was considered a laborious process until a revolution came into the pharmaceutical industry when computational software provided novel hits in drug designing. This was named CADD, which involves computer-based tools to analyze molecules on screen for drug development.

The development of a drug begins with scientific knowledge about understanding disease, receptor-targeted and active sites, *etc.* It further includes clinical trials to check safety and effectiveness against disease. This process may take up to 3-6 years and takes up a high cost. CADD, also called the In-silico method, collaborates with biologists, pharmacists, and computer engineers to bring a new drug entity quickly. CADD facilitates the drug development process as everything can be documented on screen.

Numerous drugs have credited their formation to CADD: angiotensin-converting enzyme (ACE) inhibitor captopril, carbonic anhydrase inhibitor dorzolamide, and indinavir, *etc.*

The CADD method involves 3 stages, identification of target molecule; the interaction of the protein with the active site by docking, and formation of a lead compound by passing pharmacokinetics.

The two best approaches or most common types of CADD are Ligand-based drug design (LBDD) and Structure-based drug design (SBDD). When 3-D structure of the protein is known, SBDD is used; otherwise, to deal with unknown structures, ligand-based drug design makes its way. CADD methods are dependent on bioinformatics tools, applications, and databases. Molecular dynamics simulation is the main tool in small molecule conformation, and it helps in modelling conformational changes in a biological target when bounded by active sites of other proteins. 'ab Initio method is used to suggest electronic properties of the drug candidate that impact binding affinity and help find a new drug. The other tools used in CADD are QSAR, docking, pharmacophore and visual screening, *etc.*

Structure-based drug design (SBDD) involves x-ray crystallography to take the protein's 3D structure, which is mandatory. The three-dimensional molecular structure of not less than 2.5 Å is mandatory in CADD. In Ligand-based drug design, the structure of a protein is unknown, and it focuses on other active molecules having potential against biological targets. It involves Pharmacophore models taken from other active molecules having characteristics to bind to the biological target.

The basic tool used in CADD is molecular docking. It predicts the structure of an intermolecular complex protein. The pharmacophore approach is mainly related to biological activity. It gives information about the interaction of biologically active molecules to targeted proteins along with 3D structure.

In between drug and target molecule, the extent of interaction and binding energies are also calculated by docking. It is mainly used in ligand-based drug design to identify which active site of a ligand will bind to the known structure is not well defined QSAR tool is used as in ligand-based drug design. QSAR (Quantitative structured activity relationship) gives information about the relationship between predictor and response variables by using graphical models. The main advantage of QSAR is that chemical compounds are not needed to be tested. Then CADD tools were expanded to virtual screening (lead discovery by computational screening). This is regarded as a key methodology in Computational-aided drug design.

The main advantages of computational aided drug development include reducing time and cost: it reduces half of the cost used in drug development. Also, it has the power to predict promising lead candidate's selection, which prevents time wastage on dead-end compounds.

CADD is a very significant technology as it gives information about possible derivatives on a molecular basis. It involves the analysis of 3-D structure of the protein that helps in active site identification. It helps in new compound formation by joining small fragments or functional groups.

Still, there are limitations in using computational techniques for drug development as the biological processes cannot be fully simulated on screen. Secondly, the biggest hindrance is target flexibility; docking tools make ligand flexible, but protein is fixed. This makes the actual result different from calculated values. Also, water activity must be considered in the computational analysis as in actual analysis; it alters the result.

In the future, new technologies like ADMET are being introduced to have accurate results and functional genome study for new drug targets achievements quickly (Nag & Dey, 2011).

BIG DATA ANALYSIS

Big Data

The field that uses the ways to examine comprehensively extracts facts or information, or the huge amount of complex data to be dealt with the traditional data processing application software. As reported by Forbes, almost 2.5 quintillion bytes of the data are produced every single day. This number has constantly been increasing in the following years.

Significance of Big Data Analytics

- Reduction in the Cost
- Fastened and efficient decision making
- Greater customer services.

Big data can be defined by the following three properties:

1) Volume

Due to the huge quantity of data, storage is not possible on a single machine. How can we do process our data over various machines persuading fault resistance?

2) Variety

How can data be handled originating from diverse sources that have been formatted by utilizing various formatted strategies?

3) Velocity

How can we rapidly stock and process our latest data?

Processing Techniques for Analysis of Big Data

◆   Batch processing

◆   Stream processing

Batch Processing

This processing technique is commonly used when we are dealing with the variety and volume of the data. Firstly, we store all the data needed and process it at once (high latency). The calculation of monthly payroll summaries is an application example of batch processing.

Stream Processing

This processing technique for big data analysis is used when we are focused on quick response times. In this, we process our data immediately as it is received. A common example of stream processing for identifying bank transaction is illegal or not.

Various tools can do big data analytics. MapReduce, Spark, Hadoop, Pig, Hive, Cassandra and Kafka**.**

MapReduce

When we are dealing with big data, we may end up with resources. Here, there are two possible solutions **vertical scaling** and **horizontal scaling**.

Vertical scaling can be done by adding up additional computational power (RAM, CPU), and in horizontal scaling, we distribute workload by adding further machines of the same capacity. MapReduce work on the basis of horizontal scaling, which uses clusters of computers to handle the big data. Input data is divided into different parts, and every part sends to various machines which process it and aggregate data as per specified group function.

Apache Spark

This tool has been developed as an improvement of MapReduce. Its execution speed makes it different from other tools. It is 100 times faster than that of the MapReduce tool. The programming languages used in it are R, Python, and Scala.

Used for

- Pre-processing of a huge number of data (SQL)
- Machine learning approaches are used for the analysis of data, and they process the graph networks.
- When the spark is used, big data is parallelized by resilient distributed databases (RDDs), which retrieve the lost data in case of any workers' failure.

Big Data Analytics in Genomics

The large amount of data that has been generated by sequencing, mapping, analyzing genomes leads to genomics into the domain of big data. Traditional data analysis methods may not be efficient, focusing on the demand of big data analytics in genomics. Genomics can generate a large amount of data; The human genome consists of 20-25K genes which consist of 2.3 million base pairs. This amounts to 100GB of the data, which is equal to 102,400 photos.

The various computational methods and genomics analysis techniques NextBio, Portable Genomics, Bina technologies can assist users in accessing and analyzing the sequencing data and in answering all biological queries. The concept of personalized medicines can be introduced by Big Data analytics.

In the present era, we are generating more data than we actually can process it. Thanks to the recent developments of big data in the field of artificial intelligence, which enabled machines to do the work that was impossible in past years (Ankam, 2016).

Responsible Genomic Data Sharing

The genomics culture of sharing data is robust and powerful. We are currently approaching the sign of solid assumptions of two decades in sharing genome-wide transcriptomic techniques and related metadata. This abundance of information has empowered new methodologies that depend on examining a huge collection of public data by specialists who were not associated with the collection of original data. It is likewise conceivable to test genotypes, methylation, and numerous different characteristics of a genome-wide sample, which presents significant open doors for a secondary examination. The sharing cycle has gotten murkier with verification of concept-based studies, demonstrating the possibility of interestingly recognizing a person in steadily extending sorts of definite data indexes. The risk of reidentification from the samples that are derived from humans has been expanded due to the switching of expression profiling to sequence-based from array-based. The database of Genotypes and Phenotypes serves to mediate genetic data risk of reidentification that has prompted controlled-access sharing.

Nonetheless, genotype data consisting of totaled estimates like statistics of the variant level association represent some risk that people could be reidentified. Funders, agents, and other investigators that are supporting the sharing of responsible data should consider both the pros and cons not only to members but also to the participants who could be influenced emphatically or contrarily by sharing a data set of research in various ways. In addition to moral concerns, it is imperative to consider the effect of practices of sharing data on the overall ecosystem of research. Technologies of genomic profiling are currently universally accessible and are getting broadly utilized in fields with various societies of sharing (Table **7.1**). Funders and distributers should adjust different considerations to create proper policies.

Moreover, early genomic researchers perceived the potential for high-dimensional profiling to prompt irreproducible outcomes and false discoveries if source information were not shared. Funders and distributers eventually should find a way to cultivate a piece of strong, capable information offering society to help thorough research to high-dimensional genomic profiling advances. Examiners who have shared data well increment the effect of their research, publications connected to an information archive, or constant identifier are more cited. Research scientists who produce genomics data can find a way to make that information as effective as expected under the circumstances: adding key metadata components, offering the data to the least limitations conceivable, and placing data in repositories specific to data type.

Nonetheless, making a capable culture of sharing data that quicken research is something beyond the duty of the individuals who produce data throughout their examination. Data of human study participants for controlled access, those dissecting the information should do as per the consent of members and provided study plans. Journals have to decrease to publish investigations that are not directed as per moral exploration rehearses. Funders should help moral examination in assorted populations, especially those who have set up commendable records of creating broadly reused assets (Byrd, Greene, Prasad, Jiang, & Greene, 2020).

What are Research Data?

There are many different classes and types of research data in genomics. The data can be divided by the presence of different types of biomolecules. For example, different assays are available in which some measure RNA while others measure protein, DNA, or other metabolite content. The data can also be divided with respect to measurement technology which is used to gather the data. For example, we perform different assays through different techniques like RNA profiling through sequencing or microarray. Moreover, a sample can be derived from various sources, including any organism, human tissue, cell line, or a specific population of specific origin.

The major purpose of responsible data sharing is to collect or consider genetic data that can gene profiling or profile the gene products for most of the organism's genome. The derived data, an intermediate of research data, and the raw data obtained from the experiments are also important in sharing genomic data. For example, the read files in FASTQ format obtained from RNA sequencing experiments are considered raw data. Likewise, the gene expression analysis is considered intermediate data, and figures, tables, or other statistical analysis of gene expression data is considered a finding. So, there could be many representations of intermediate data which lie between findings and raw data.

Another example is the sequencing of normal and paired tumor samples that characterize somatic and germline polymorphisms. The results from these experiments generate FASTQ files for each normal and tumor sample, files of variant cell format, files of mutation annotation format, and the results summary and figures. In this scenario, there can be multiple intermediate files between findings and the raw data. In responsible genome data sharing, the recommendations are provided for how researchers select the materials from raw data to findings that should be recorded and how this data can be best shared.

Now, derived data is thought to be a model generated by the machine-learning method when applied to genomics. Investigators can now process their data according to their studies, analyze them through neural networks, and then use these models to understand a particular disease's methodology or develop understanding. Machine-learning algorithms provide the basis to analyze every type of data, either raw data or any finding. These algorithms can help us to maintain distinctions between raw data, intermediate data, and findings. With the help of these data models and those that will develop in the future can be applied to generate a better sharing plan (Villanueva, Cook-Deegan, Robinson, McGuire, & Majumder, 2019).

Table. 7.1. Types of Genomic data and its associated risk.

Type of Data	Usual Level of Risk	Sharing to Lessen the Risk
Reads of RNA-seq of model organisms	None	Public Access
Whole-genome sequencing profiling of Species that are endangered	Generally none, even though area metadata could place species at risk	Public data but controlled-access metadata
Reads of RNA-seq of samples of human tissue	High	Estimates of expression of Public gene
Reads of Whole-exome sequencing of samples of cancerous tissue	High	Public access for data of somatic variant, but controlled access for data of variant of germline
Exome sequencing of samples of non-cancerous tissue of human	High	Public rundown level data collected across numerous people
High-density DNA methylation array of human tissue	High	Eliminate data from tests that contain normal variations before open sharing Controlled admittance for full data collection

CONCLUSION

This chapter presents a detailed description of the role of bioinformatics in the post-genomic era, Digital Sequence Repositories, From the world of 1D to the era of 3D (Protein Structures), functional genomics, sequence alignment, phylogeny and concept of gene evolution, sequence and structural alignment, genomics of personalized medicine. Moreover, it also summarized the information regarding CADD: Computer-Aided Drug Designing, big data analysis, and responsible genomic data sharing.

REFERENCES

Ankam, V. (2016). *Big Data Analytics.* Packt Publishing Ltd.

Byrd, J.B., Greene, A.C., Prasad, D.V., Jiang, X., Greene, C.S. (2020). Responsible, practical genomic data sharing that accelerates research. *Nat. Rev. Genet., 21*(10), 615-629.
http://dx.doi.org/10.1038/s41576-020-0257-5 PMID: 32694666

Fragai, M., Luchinat, C., Parigi, G. (2006). "Four-dimensional" protein structures: examples from metalloproteins. *Acc. Chem. Res., 39*(12), 909-917.
http://dx.doi.org/10.1021/ar050103s PMID: 17176029

Ginsburg, G.S., Willard, H.F. (2009). Genomic and personalized medicine: foundations and applications. *Transl. Res., 154*(6), 277-287.
http://dx.doi.org/10.1016/j.trsl.2009.09.005 PMID: 19931193

Nag, A., Dey, B. (2011). *Computer-Aided Drug Design and Delivery Systems.* McGraw-Hill Education.

Pevsner, J. (2015). *Bioinformatics and functional genomics.* John Wiley & Sons.

Rosenberg, M.S. (2009). *Sequence Alignment: Methods, Models, Concepts, and Strategies.* Univ of California Press.
http://dx.doi.org/10.1525/9780520943742

Shanker, A. (2018). *Bioinformatics: Sequences, Structures, Phylogeny.* Springer.
http://dx.doi.org/10.1007/978-981-13-1562-6

Torshin, I.Y. (2006). *Bioinformatics in the Post-genomic era: the Role of Biophysics.* Nova Publishers.

Villanueva, A.G., Cook-Deegan, R., Robinson, J.O., McGuire, A.L., Majumder, M.A. (2019). Genomic data-sharing practices. *J. Law Med. Ethics, 47*(1), 31-40.
http://dx.doi.org/10.1177/1073110519840482 PMID: 30994063

Ziemann, M., Kaspi, A., El-Osta, A. (2019). Digital expression explorer 2: a repository of uniformly processed RNA sequencing data. *Gigascience, 8*(4)giz022
http://dx.doi.org/10.1093/gigascience/giz022 PMID: 30942868

CHAPTER 8

Human Genome Project

Abstract: This chapter proposes a brief description of Human genome Project; Budget and Goals of Human Genome Project; Human Genome Project Timeline; Science behind Human Genome Project; Advances based on Human Genome Project; Human Genome Project Impact; Influence of human genome project on biological and technological progress; Medicinal Impact; Effect on law and the sociologies; Technical Aspects of Human Genome Project; Mapping Strategies; Sequencing Strategies; Findings of Human genome Project; Post Human Genome Project era: What is to come.

Keywords: Human genome Project, Mapping, Sequencing, Post Human Genome Project.

INTRODUCTION

The Human Genome Project has been greeted as a significant achievement throughout science and as a venture whose fruition would change the act of medication. The Human Genome Project is as important as sending man on moon or isolating an atom. Many have argued that the Human Genome Project addresses a perspective change in science, significantly changing how biomedical examinations are finished.

Gene hunting offered a path to a considerably more efficient and data-driven methodology. Data-driven depicts a type of exploration that, instead of depending on the cognizant planning of investigations to test theories, depends on gathering a lot of information that can be deliberately dissected, perused, and fished with the assistance of software programs searching for affiliations and patterns. Enormous measures of sequencing data were collected in huge scope data sets, considering a more exhaustive examination of living frameworks. Human Genome Project has released a cascade of research papers of stunning extents, covering different sorts (research articles, books, remarks, interviews, and so forth). An international collaboration involving many countries started the human genome project to sequence the human genome completely. The project was started in 1990 and completed in 2003, suggesting three billion base pairs in one complete human

Maryam Javed, Asif Nadeem & Faiz-ul Hassan

genome. The Human Genome Project was also expected to enhance the advancements to decipher and dissect sequences of the genome to recognize all the genes encoded in humans' DNA and resolve the social, legitimate, and moral ramifications that arise from characterizing the entire human genome sequencing (Sawicki, Samara, Hurwitz, & Passaro Jr, 1993).

BUDGET OF HUMAN GENOME PROJECT

When the human genome project was broadcasted, numerous researchers contended that it was superfluous to sequence the whole genome as it included a ton of cash. Along these lines, explicit goals were created, including the most effective methods that would cost less. The underlying appraisal for the Human Genome Project was 200 million US$ every year for a time of 15 years. Quick advancement was made in sequencing procedures and mapping strategies which prepared for the finish of the Human genome project ahead of its objective timetable 2005. Consequently, the cost of the project was not as much as what had been normal. The spending plan of the Human Genome Project was roughly 2.7 billion US$ rather than 3.0 billion US$. The USA gave around 50% of the asset, and the rest came from different nations, for the most part, the UK, France, Germany, Japan, Australia, and Canada. In 1988, the Office of Technology Assessment endorsed assets for the HGP through the National Institute of Health and DOE, which were 17.3 million US$ and 11.8 million US$, separately (Ingelman-Sundberg, 2005).

GOALS OF HUMAN GENOME PROJECT

The thoughts behind the HGP were to become familiar with the hereditary mechanism that controls the development of the human from the zygote, its connections with the climate, hereditary problems, the process of aging, and so on. It is difficult to see all these in a solitary trial. Hence, determining objectives for the Human Genome Project were set up, which covered not just the issues connected straightforwardly to the human but also tended to other related issues. When the Human Genome Project began, accessible sequencing innovations and mapping strategies were not progressed enough to sequence 3 billion bases of the genome of humans in the predetermined period. They were excessive, slow, and not precise. Accordingly, explicit objectives were set to expand the proficiency. The aims of the Human Genome project were isolated into three five-year plans before continuing to the following five-year plan; the advancement and obstacles were examined, and appropriate changes were made. The particular targets of the Human Genome Project were together set up by the National Institute of Health and the Department of Energy for three five-year time spans, 1991–95 (Collins *et al.*, 1998).

The particular objectives of the Human Genome Project are to:

1. To set a map of the high resolution of the human genome utilizing genetic and physical mapping techniques.

2. To inquire about the nucleotide arrangement order in all 22 autosomes and the two sex chromosomes (X and Y).

3. Development of high throughput sequencing technology.

4. To become familiar with the genomic sequences of model life forms to test the attainability of various sequencing and mapping strategies.

5. Create PC software to store the sequence information and access the stored sequence data for different purposes.

6. DNA sequence annotation dependent on its sequence content.

7. To address moral, social, and legitimate issues that may emerge relating to human genome sequencing and its utilization.

8. Recognize all the assessed 80,000 genes in the DNA of a human.

HUMAN GENOME PROJECT TIMELINE

Preceding the Project of Human Genome, numerous individual researchers had addressed the various base sequences of the human gene. Since most of the genome of humans stayed neglected and the scientists felt the need and importance of having fundamental data of the human genome at hand, they were starting to explore the paths to reveal the data more swiftly. As billions of dollars are required for human genome projects that many biomedical scientists, political parties had supported, conventional biomedical researchers and ethicists got engaged in enthusiastic discussions over the merits, demerits, potential benefits, and the approximate cost in terms of sequencing the whole genome of human in a single endeavor. Despite arguments, 1990 was the beginning of the project human genome lead by Francis Collins, an American geneticist, supported by the National Institute of Health and United States Energy Department. Soon, many scientists around the globe started to join the effort. Moreover, several technical advances have been made in sequencing by advancing the hardware and software of computers used to examine and track the resulting outcome. Hence, enabling the swift progress of the project.

One of the forces determining the speed of human genome project discovery were the advances in technology. In 1998, an American Scientist who runs the private sector lab known as Celera genomics and the former Scientist of the national institute of Health J. Craig Venter started sabotaging and potentially competing with the human genome project. The competition was totally gaining full control over the intellectual patents on the human genome sequence that was depicted as a treasure trove in pharmaceutics. The rivalry of control over the project between NIH and Celera genomics ended as soon they became part of the forces that ran the project to speed the completion of the rough draft of the genome of humans. However, their financial and legal issues remain unexplored. In June 2000, Collins and Venter declared the completion of the rough draft of the genome of humans. The rough draft of the genome of humans was further refined, examined, and expanded for the following three years and in April 2003, the human genome project was announced complete (Roberts, 2001).

SCIENCE BEHIND HUMAN GENOME PROJECT

It is important first to determine the pillars of science that were the field of classical genetics and molecular biology upon which the human genome project was based to appreciate the implications, challenges, and directions of the human genome project. 1800 was the beginning year of classical genetics introduced by Gregor Mendel, an Austrian botanist, who explained the basic genetic laws in his research of the garden pea. Basic rules of genetics explained by Mendel's were further extended in 2oth century, just as the molecular biologist initiated research utilizing the model life forms like Drosophila melanogaster. Studies on the model life forms help better understand the key concepts of genetics and complexities observed in the transmission of genetics. Filed molecular biology evolved soon to realize DNA and RNA as the hereditary material in all life forms.

Physically, the gene combines four different types of nucleotides (A, G, C and T). The specific arrangement and number of these nucleotides are genetic information that first encode RNA and protein as final products. For some of the genes, RNA is the final product. These final products (RNA or protein) play structural and functional roles in a living cell.

Molecular genetics elucidates humans acquired characteristics. Most of the attributes of an organism are the result of inheritance and ecological impacts. Closely linked genes on the same chromosome are inherited together with the least chance of assortment. Mitochondrial genes are solely inherited from the mother, and Y-chromosome is transferred from father to son only. Human Genome Project has estimated 20,000 to 25,000 genes in the human genome (Cantor & Smith, 2004); (Croce, 2015).

ADVANCES BASED ON HUMAN GENOME PROJECT

Development in hereditary studies and genomics keep on arising. Two significant advances incorporate the International HapMap Project and the inception of comparative genomics studies, two of which have been made conceivable by the accessibility of data sets of human genomic sequences and the accessibility of databases of genomic sequences of a large number of different species. The International HapMap Project is a cooperative exertion between Japan, United Kingdom, Canada, China, Nigeria, and the United States. The objective is to distinguish and inventory hereditary similitudes and contrasts between people addressing four significant human populaces got from the landmasses of Africa, Europe, and Asia. The ID of variations in genetics consider polymorphisms that exist in sequences of DNA among populaces permit specialists to characterize haplotypes. These markers recognize explicit regions of DNA in the genome of humans. Affiliation investigations of the predominance of these haplotypes in control and patient populaces can be utilized to help recognize possible functional hereditary differences that incline a person toward sickness or that may shield a person from infection. Essentially, linkage investigations of the inheritance of these haplotypes in families influenced by a known hereditary trait can likewise assist with pinpointing the particular gene or genes that underlie or change that characteristic. Affiliation and linkage examination have empowered the distinguishing proof of various genes and their transformer.

As compared to the International HapMap Project, which analyzes genomics sequences inside one species, comparative genomics is the investigation of likenesses and contrasts between various species. Lately, an incredible number of full or practically full genomic sequences from various species have been resolved and stored in open data sets, for example, NIH's Entrez Genome data set. By looking at these sequences, regularly utilizing a programming software apparatus called BLAST, specialists can distinguish levels of likeness and difference between the genes and genomes of related or different species. The consequences of these investigations have enlightened the development of species and genomes. Such examinations have additionally assisted with causing to notice profoundly conserved areas of non-coding DNA sequences that were initially thought to be nonfunctional because they don't contain base sequences that are converted into protein. Nonetheless, some non-coding areas of DNA have been profoundly conserved and may assume key parts in human advancement (Collins, 1998);(Wilson & Nicholls, 2015).

HUMAN GENOME PROJECT IMPACT

Influence of Human Genome Project on Biological and Technological Progress

Sequencing of humanoid genetic material led to thorough detection and classification of a 'parts list' largely comprised of genes, *via* deduction maximum of the human proteins, together with additional essential elements including regulatory RNAs that do not code for proteins. It is crucial to learn about the basic elements, how they connect, the dynamics involved, and how all these lead to the ultimate function to understand the subsequent complicated biological system. The parts list has proven indispensable for the advent of 'systems biology', which has changed the approach to perceiving biology and medicinal sciences. For instance, the Encyclopedia of DNA Elements (ENCODE) Project was initiated by National Institute of Health (NIH) in 2003 and it intends to find out and understand the functional elements of the human code. The ENCODE Project Consortium has generated volumes of valuable data relevant to the regulatory networks that control gene expression *via* multiple approaches, most of which were based on 2nd generation sequencing technologies. Large sets of data like those generated by ENCODE gave rise to thought-provoking questions about the functionality of human genetic material. How can one differentiate a real biological hit from the unavoidable clatter that results from large sets of data? To what level is the functionality of an independent genomic element only applicable in a precise context (such as mRNAs and regulatory networks restricted to operate in embryogenesis)? Obviously, a lot is still to be accomplished before the functionality for the inadequately noted protein-encoding genes are decoded, let alone that of the enormous parts of non-coding fractions of the transcribed genetic material. However, the difference between noise and signal is a crucial point to be considered.

The Human Genome Project (HGP) also contributed to the advent of 'proteomics', a field engaged in identifying and quantifying proteins located inside distinct physiological partitions, for instance, in blood, cell organelle, or organ itself. In their diverse roles such as molecular machines, signaling devices, or structure-based elements, proteins form the basis of exact cell functions of parts list in a being's genetic code. The HGP aids in the application of a principle analytical technique that is mass spectrometry, *via* making reference genome available. Hence, the expected masses of tryptic short chains of amino acids in the human proteome are fundamental for mass spectrometry-based proteomic analysis. The access to the vast proteomic data due to mass spectrometry has steered innovative applications like 'Targeted Proteomics'. However, proteomics demands complex computational techniques such as the Trans-Proteomic Pipeline and PeptideAtlas. The HGP has transformed our understanding of evolution. Since the HGP has been completed, more than 4,000 drafts of genomic sequences have been generated, most of them from microbial species but include about 183 eukaryotic sequences as well. This data presents an understanding of how distinct living creatures, from microorganisms to human beings, are interrelated on the ancestral 'tree of life' – evidently establishing that all existing species originate from a common ancestral root. Longstanding questions of scientific interest that held implications for biological and medical sciences can now be approached. How are novel genes introduced? What function do fractions of protected sequences found across all metazoa perform? To what extent is the large scale organization of genes across species conserved, and what steers local and universal rearrangement of the genome? Which parts of the genetic material seem to be unaffected/susceptible to mutation or are more prone to recombination? How does the evolution of regulatory networks take place and change the expression of genes? The latter query is now fascinating since many hominids or primates now have their genomes sequenced orin-process and can un-leash the data related to the evolution of diverse human features. The genome sequence of Neanderthal has led to an implication of interest related to human evolution; some percentage of Neanderthal genome and thus the subsequently coded genes are intermingled in the human genetic code, which proposes the presence of inter-reproduction between the 2 species during divergence.

The HGP has nurtured ties between cross disciplines as a force that steered the advancement of complex mathematical and computational approaches to deal with data and brought mathematicians, computer scientists, theoretical physicists, and engineers together with biologists. It is vital to consider that HGP made the notion of making scientific data accessible to the community more popular *via* manageable databases, for instance, the UCSC Genome Browser and GenBank. Furthermore, the HGP endorsed the concept of open-source software, and the source code was made available to be edited to fill the room for improvement and expand outreach. The open-source operating system such as that of Linux and the resultant community it has outreached has revealed the potential of this approach. The idea of accessible data is of utmost magnitude for the triumph and culture of the biological arena in the future as 'democratization of data' holds a vital role in drawing the available aptitude to concentrate on the challenges that come with biological systems and related innate complications. This is even more crucial in the field of medicines, as researchers require access to the data cloud made obtainable from every individual being in order to excavate for future prophetic therapeutics. This attempt has the potential to impact public health over generations.

The HGP is the foremost instance of 'big science' in the biological field and has exhibited both the necessity and potential of this approach for handling its integrated technological and biological aspirations. The HGP had a well-defined set of motivated ambitions and impressive plans to accomplish them; selective funded investigators in organized consortia or centers; a clear commitment to resource release/public data as well as a prerequisite for substantial resources to sustain the infrastructure of the project and novel technological advancement. Big data science and small-scale independent investigator-directed research are balanced. The former creates funds that are deemed fundamental for all scientists. The latter supplements comprehensive scientific demonstration of a specific query and enhances analytical power and depth to the data generated by the big data science. Uncountable layers of complexity exist in biological and medical sciences, and such projects based on big data sciences are critical to solving these layers in an integrative and detailed manner.

The HGP has aided biological and medical sciences by deciphering the complete sequence of human genetic code, developing high throughput sequence-based technologies, providing reference sequences from model organisms, and exploring social and ethical concerns related to such advanced technologies. The project benefited from corresponding economic advantage and the directed force of a global consortium with chosen participants, which reduced effort to a much more effective level than would have been achievable if the genomes were sequences on a gene-by-gene basis at a pilot scale. Moreover, it is important to note that HGP had a vast aptitude for economic gains that appealed to the government to assist it. The economic influence of HGP was published in a report by the Battelle Institute. The return on the initial investment of about $3.5 billion was approx.—$ 800 billion in accordance with a report – an astounding return on investment.

Even at present, there is an inclination to tract support from big data science as budgets drop and concentrate available funds on small-scale science. But it is a drastic misinterpretation. In the face of HGP, there are other projects based on the generation of biological resources and studies related to physiological intricacy that need a big science strategy, such as the ENCODE project, the HapMap Project to catalogue human genetic variation, the European Commission's Human Brain Project, and the Human Proteome Project as well as another brain-mapping project announced by President Obama. Like the HGP, substantial returns on the investment are to be expected from other projects that adhere to the big science approach that is at present under consideration if they are accomplished well. However, it must be underlined that caution must be taken to decide such projects profound of significance. It is clear that the funders must maintain a mixed and balanced portfolio of large and small-scale science – and two are synergistic in nature.

Last but not least, the HGP has lit the creativities of talented and remarkable researchers - Sydney Brenner, John Sulston, Jim Watson, Bob Waterston, and Eric Lander, to name a few. Hence, each argument stated by the antagonists of HGP in its opposition turned out to be incorrect. The HGP is a fascinating instance of a vital paradigm transformation in biology that was resisted initially but ultimately turned out to be more phenomenal than predicted by even the most enthusiastic supporters (Uddhav & Ketan, 1998).

Medicinal Impact

The complete sequencing of the human genome and its availability on public databases presents a defining event for the scientific community and benefits society. Following the completion of HGP, a number of big science ventures started with a specific aim to improve the current knowledge of genetic variations in humans and their correlation to health and disease. Such projects include the HapMap Project that was directed towards the identification of haplotype blocks that are composed of frequent variations in the form of SNPs in various human populaces, and then came the 1000 Genomes project that is still in process and aims to catalogue rare and common structural and single nucleotide variation in numerous populaces. The information generated *via* these projects has promoted small scope genome wide association studies (GWAS), which connects causative genetic variations to disease susceptibility of variable numerical importance on the basis of health and disease group comparisons. Furthermore, these achievements, in turn, have empowered the accurate and sensitive diagnosis of diseases even before the representation of specific clinical symptoms. Moreover, Linkage and genetic association studies put the additional impact which brings good results for both common and rare diseases.

From 2005, more than 1,350 Genome-wide association studies have been reported. Though GWAS provides clues to the genome's location to search for causative variants, the outcomes can be tough to decode as the size of the experimental sample could be too small, the actual causative variant could be rare, or the phenotype of the disease could not be well stratified. Furthermore, more than 50% GWAS hits are not in the coding part of the genetic code. At present, there are no efficient methods to resolve in a simple manner if these hits indicate the incorrect working of regulatory units. It is of concern as to what portion of the thousands of GWAS hits are true indicators and what is clatter. However, pedigree-based whole-genome sequencing (WGS) proposes a significant alternative approach to identifying probable disease causative variants. Though only a number of personal genomes were completely sequenced 5 years ago, now these are thousands of exomes and whole-genome sequences that will soon be tens of thousands, and later millions, that have been revealed to uncover the causative disease-related variants and, on a greater scale for the establishment of distinct connections amongst variations in sequence and specific observable phenotype. For instance, The Cancer Genome Atlas and the International Cancer Genome Consortium are working on large genomic datasets and a number of cancer types, promising to make the resultant resources accessible to the rest of the scientific arena.

It is assumed that individual genomic sequencing would shortly have a greater role in the medical field. In an ideal case, consumers, users, or patients would utilize this data to better their health situation by gaining the benefit of therapeutic or preventive approaches established to be suitable for potential or actual medical scenarios as recommended by the individual genomic sequence. Doctors will also have to educate themselves to advise the patients that present consumer genomic data in their appointments, which will become a usual occurrence in the near future. System biology has brought a revolution in our healthcare system. P4 medicines are one of the achievements, similarly many of the diagnostic approaches have been developed. Pharmacogenomics has found variation in over 70 genes associated with the inefficiency to metabolize the drugs (Williams & Hayward, 2001).

Third, in some patients, tumorous cancer-related mutations, if discovered, can be offset *via* therapeutic intervention. Lastly, some systems biology methods to blood-based proteins diagnostics have produced influential diagnostic markers for human ailments like lung cancer and hepatitis.

These latter instances showcase a transformation in blood-based diagnostic technologies that can lead to detection of diseases at an earlier stage, the potential to follow the progression of the disease and reactions to therapy, and capability for the stratification of a particular disease group such as breast cancer into its various subgroups to match against accurate therapy. In the future, a time can be envisioned when there will be analytical techniques to diminish this huge data to straightforward hypotheses in order to improve health and lessen the disease rate for each individual. One of the greatest influences of the human genome project was the introduction of the concept of personalized medicine was built on the idea that the human genome sequence will help doctors and healthcare experts to give more appropriate and effective medicine to patients. No doubt, proceeding with propels in DNA sequencing innovation guarantee to bring down the expense of sequencing a person's whole genome to that of other, moderately cheap, diagnostic tests (Collins & McKusick, 2001).

Effect on Law and the Sociologies

The human genome project led the researchers to have thoughts about the social implications of their research. Surely, it committed 5% of its budget to think about the legitimate and moral issues of understanding the human genome sequence and procurement. That cycle proceeds as various cultural issues emerge, for example, hereditary security, expected discrimination, equity in distributing the advantages from genomic sequencing, human subject assurances, hereditary determinism, character governmental issues, and the philosophical idea of being individuals who are inherently associated with nature. Strikingly, we have gained from the human genome project that there are no specific race-explicit genes among people. Maybe it can happen when a person's genome uncovers their tribal heredity, an element of the relocations and interbreeding among populace gatherings.

The Human Genome Project influences fields past biomedical science in manners that are both unmistakable and significant. For instance, information of genomic sequences of humans, investigated through a framework called CODIS abbreviated as Combined DNA Index System, has changed the field of legal sciences, empowering positive ID of people from very small examples of natural substances, for example, spit on the envelope seal, a couple of hairs, or a spot of semen or dried blood. To be sure, prodded by high paces of recidivism (the inclination of a formerly indicted criminal to re-visitation of earlier criminal conduct instead of discipline or detainment), a few governments have even initiated the arrangement of banking DNA tests from all sentenced crooks to encourage the distinguishing proof of culprits of future violations. While politically questionable, this approach has demonstrated exceptional viability. By a similar token, blameless people have been absolved based on DNA proof, here and there a very long time after unfair feelings for wrongdoings they didn't submit. Comparative DNA sequences examinations of tests addressing noticeable current populaces of people have altered the field of human sciences. For instance, by following sequence variations of DNA existing on mitochondrial DNA, which is maternally acquired, and on the Y chromosome, which is in a fatherly way acquired, molecular anthropologists have affirmed Africa as the support of the advanced human species called Homo sapiens, and have recognized the rushes of human migration that arose out of Africa in the course of the most recent 60,000 years to populate different mainland of the world (Cowan, Kopnisky, & Hyman, 2002);(Ellsworth, Hallman, & Boerwinkle, 1997).

Technical Aspects of the HGP

The way toward deciding the human genome includes first mapping or portraying the chromosomes. This is known as a physical map. Sequencing is the next phase, or deciding the DNA bases order on a chromosome. These are known as genetic maps.

Mapping Strategies

Maps are required in order to sequence the genome of humans. Physical maps are a progression of overlapping bits of DNA disengaged in microbes. Physical maps are utilized to depict the chemical attributes of DNA. Mapping includes isolating the chromosomes into pieces that can be proliferated and portrayed and afterward ordering them to compare to their individual chromosomal areas. Hereditary markers are important for mapping the genome. Markers are any acquired physical or molecular attribute that is distinctive among people of a populace. This depicts differences in sequences of DNA spots that limited restriction enzymes can cut. To be helpful in mapping, markers should be polymorphic or have more than one structure among people to be perceivable in examinations. Another marker is Variable Numbers of Tandem Repeats, which are little segments of repeating DNA. The variable number of tandem repeats is pervasive in the DNA of humans and can exist in wide differences of numbers. This inconstancy gives people special VNTR locales. This is the application behind tackling criminal cases with blood tests. A hereditary map shows the general areas of these particular markers on chromosomes.

Utilized in RFLP markers are limitation chemicals. These enzymes perceive short DNA sequences and cut them at an exact spot. Since researchers have described many diverse restriction enzymes, DNA can be cut into various sections. These parts are the pieces of DNA utilized in physical maps. Various kinds of physical maps exist. Physical maps of low resolution incorporate chromosomal maps that depend on particular patterns of banding of stained chromosomes. Physical maps of high-resolution address sets of DNA sections cleaved by restriction enzyme and put in order as previously depicted (Council, 1988);(Balmer, 1996).

Sequencing Strategies

In order to sequence DNA, it should be first amplified or expanded in amount. Cloning and Polymerase Chain Reactions are the two kinds of DNA amplification methods. Cloning includes the propagation of fragments of DNA into a foreign host. Restriction enzymes isolate the fragments of DNA and then join with a vector and afterward replicate alongside the DNA of the vector's cell. Vectors ordinarily utilized are viruses, microscopic organisms, and cells of yeast. Cloning gives a limitless measure of DNA for exploratory investigation.

PCR amplifies DNA hundreds of millions of times, an assignment that would have required days with recombinant DNA technology. PCR is significant on the grounds that the response is exceptionally explicit, effectively computerized, and fit for amplifying an extremely modest quantity of DNA. Thus, PCR significantly affects clinical medication, the determination of genetic illnesses, the science of forensic, and transformation biology.

PCR is a cycle through which a specific polymerase enzyme orchestrates a complementary DNA strand to a different given strand of DNA in a combination of DNA bases and DNA pieces. The blend is warmed, isolating the two strands in a double-stranded molecule of DNA. The blend is then cooled, and through the activity of the enzyme polymerase, the DNA parts in the mixture find and stick to the sequences that are complementary on the currently isolated strands. The outcome is two double helix strands from one double helix strand. Repeated cycles of cooling and heating in PCR machines amplify the DNA that is targeted dramatically. In less than ninety minutes, PCR cycles can enhance DNA by a millionfold (Waterston, Lander, & Sulston, 2002).

Since the DNA has been amplified, sequencing can start. Two fundamental methodologies are sequencing by the Maxam-Gilbert approach and Sanger sequencing. The two techniques are fruitful on the grounds that gel electrophoresis can deliver high-resolution isolation of DNA particles. Electrophoresis is the way toward utilizing gels with stained DNA and afterward isolating those DNA parts as per size by the utilization of electric flow through the gel. Indeed, even sections that have just one single, distinctive nucleotide can be isolated. Practically the entirety of the means in both of these sequences are presently computerized.

The chemical degradation method is the other name of Maxam-Gilbert sequencing, which separates DNA at explicit bases utilizing synthetic chemicals. The outcome is fragments of different lengths. A refinement to this technique known as multiplex sequencing empowers researchers to dissect roughly 40 clones on a solitary DNA sequencing gel.

Sanger sequencing, additionally called the chain termination or dideoxy technique, utilizes enzymes to integrate DNA of changing length in four unique responses, halting the replication at positions involved by one of the four bases, and afterward deciding the subsequent piece lengths. A significant objective of the Human Genome Project is to create robotized sequencing innovation that can precisely sequence in excess of 100,000 bases for each day. Explicit centers incorporate creating sequencing and location conspires that are quicker, more delicate, precise, and prudent. In 1991, PC innovation entered the sequencing cycle at Oak Ridge National Laboratory, where a man-made AI program called GRAIL was tried (Venter *et al.*, 2001).

Findings of Human Genome Project

1. It was discovered that just ~2 percent of the human genome is coding DNA. The total number of genes (20,000–25,000) is far less than foreseen (100,000) and nearly not exactly the previous expectation from a rough version (31,000). It was discovered that 2–3 proteins are encoded by one gene through alternate splicing.

2. Human proteome is times bigger than that of a fruit fly or a worm. It was observed that the genome has fewer exons per transcript (4.7 exons/ transcript) and small ORF (847 amino acids) than prior forecasts.

3. Human consist of 2-3 times more genes than that of fruit fly containing 13,500 genes and the worm containing 19,000.

4. Comparison of the whole genome of humans with other model organic entities demonstrated that many of the genes of humans share a sequence likeness with genes found in different organic entities. These are called orthologous genes and are therapeutically significant. This will help understand the functions of the genes in other model creatures.

5. Gene expression regulation is a significant angle in eukaryotes. A significant number of genes encode transcription factors associated with enhancers and silencers to give the adaptable expression of the gene.

6. Genes are not disseminated similarly all through the chromosome; a specific chromosome contains gene-rich and gene-poor regions. All of the chromosomes don't have the same number of genes; for instance, the human chromosome 19 has the biggest gene number while chromosome 5 has the least.

7. Phenomena of segmented duplication are seen in all chromosomes. Segmental duplications are areas on chromosomes with more than 1 kbp length, having greater than 90% sequence similarity. It has been observed that gene duplication is more common in humans than in mice, and chromosome Y has 25 percent of its length, segmental duplication.

8. Repetitive DNA represents a significant piece of the genome that fluctuates from short to long repeats. Around 85 percent of the heterochromatic locale isn't sequenced at this point, on the grounds that these fragments of DNA are hard to keep up as clones, and furthermore, in the wake of sequencing, it is hard to assemble them all together.

9. In the human genome, the total length of coding sequences is around 34 Mbp, and it represents about 1.2 percent of the euchromatic area, and 0.7 percent of sequences of the euchromatic region are untranslated.

10. The present catalogue of the human genome represents 2,287 gene loci containing 19,438 genes that are known to have been distinguished, and 2188 genes have been anticipated.

Although the Human Genome Project's definitive objective was to sequence the whole human genome, this was not accomplished. This is because of the way that the chromosomes are comprised of two sorts of sequences: gene-rich euchromatic region and the heterochromatin of telomere and centromere, which is rich in repetitive sequences and is bare from genes (Mourier, Hansen, Willerslev, & Arctander, 2001).

POST-HUMAN GENOME PROJECT ERA: WHAT IS TO COME?

It was a big and difficult challenge to understand the full human genome. For instance, so far, at any rate, 5% of the human genome has not been effectively sequenced for some reasons, which made it difficult to find eukaryotic islands in heterochromatin, variation in GC content, and copy number variations. Furthermore, there are profoundly many conserved regions present in the genome of humans that are not characterized yet, which means their function is not established. Yet, they are probably considered as regulatory regions. However, why they ought to be firmly conserved over a large portion of a billion years of development stays a mystery.

There will be more advancements keep on in genome investigation. Creating improved logical strategies to recognize organic data in genomes and unravel what this data identifies with practically and developmentally will be significant. Building up the capacity to quickly analyze total human genomes with respect to significant genetic variations is fundamental. It is additionally fundamental to create programming that can precisely crease genome-anticipated proteins into three dimensions so their native function can be anticipated from structure homology. While we have generally gotten capable of determining stable and static genome groupings, we are yet figuring out how to quantify and decipher the powerful impacts of the genome, gene expression profiling, function, and regulation of non-coding RNAs, proteins, and metabolites, and different results of encoded data.

With its emphasis on building up the technology and innovation to specify sections, the human genome project was a basic idea for launching the science of systems biology, with its associative spotlight on high-throughput 'omics' information and the concept of big data in biological sciences. The science of systems biology starts with the sections of information on different biological elements, including DNA, RNA, proteins, and metabolites. The objectives of systems biology are thorough yet open-finished on the grounds that, as seen with the human genome project, the field is encountering the implantation of skilled researchers applying multidisciplinary ways to deal with an assortment of issues. A central component of systems biology, from our perspective, is to incorporate a wide range of kinds of natural data to make the 'biological networks' - perceiving that organizations work at the genomic, the sub-atomic, the cell, the organ, and the informal community levels and that these are consistently coordinated in the living individuals. Coordinating this information permits the formation of prescient and significant models for specific sorts of living entities and individual patients. These objectives require growing new kinds of high-throughput omics innovations and ever progressively incredible scientific analytical tools.

The human genome project mixed a mechanical limit into science that has brought about gigantic expansions in the scope of exploration for both small and big science. Analyses that were undoable 20 years back are presently standard because of the expansion of scholastic and academic wet lab and bioinformatics assets outfitted towards encouraging scientific research. Specifically, fast expansions in high-throughput next-generation sequencing strategies with their corresponded diminish in the expense of sequencing have brought about an extraordinary abundance of open genomic and transcriptional succession information for plant, microbial and animal genomes. This information thus empowered small and large practical investigational studies that catalyze and improve further research and exploration when the outcomes are given in freely available public databases.

Another project which is relative to the human genome project is the Human Proteome Project, which is starting to assemble energy, despite the fact that it is still ineffectively funded. This energizing undertaking can possibly be immensely advantageous to science. The Human Proteome Project expects to make assays for all model and human proteins, including the bunch protein isoforms created from the RNA splicing and engineering of protein-encoding genes, modification of produced and mature proteins, and their processing handling inside the cell. This project, along with HGP, likewise means to pioneer advancements that will accomplish a few objectives, including empowering ELISA tests based on microfluidics for early and sensitive detection like detection from a single drop of blood, single-cell proteomics, creating protein-capturing agents that are simple, small, stable and can target only specific epitopes to avoid cross-reactivity and to build up the algorithms that will empower the standard scientist to break down the gigantic measures of proteomics information that are starting to emerge from human and different other organisms (Gibbs, 2020).

More current and novel methodologies of DNA sequencing will infer that how changes accumulate in the genome. More advanced sequencing techniques like nanopore sequencing will use nano-channels and nanopores along with the use of electric signals for sequencing of DNA molecules with greater read lengths nearly from 10000 to 100,000 bases. These properties of high-throughput next-generation sequencing will tackle numerous current issues with human genome arrangements. Initially, it was impossible to assemble and align sequences of the human genome with short-read sequencing as *de novo* human genome sequences were first compared through BLAST using a reference genome which itself was not fully accurate due to the presence of genetic variation and mutations. Due to these limitations of short-read sequencing, it was very difficult and challenging to identify and sequence the mutations and other structural variations in the genome of humans and other organisms. Long-read sequencing, also known as -generation sequencing, has shown its potential to profile the human genome and transcriptome at the cell level and their *de novo* assembly because of its ability to sequence the long sequences of DNA around 1 million base pairs at a time. Long-read sequencing also removes the need to digest DNA fragments and then their amplification which is generally required by other sequencing methodologies. Moreover, this new third-generation sequencing methodology is also involved in identifying the epigenetic modifications of the human genome because it is progressively evident that these epigenetic modifications assume significant parts in regulating gene expression. All of these sequencing advancements make the analysis of single-cell possible; thus, now we can identify the genetic and epigenetic marks. When will these third-generation technologies become mature enough to address the remaining questions of the human genome project and its related inquiries is the real challenge. From all the information, it has been concluded that the human genome project has opened many avenues in biology, biotechnology, medicine, now in computational biology, and we are continuously exploring it (Iscovich, 1998).

CONCLUSION

This chapter presents a detailed description of the human genome project, budget, and goals of the human genome project, human genome project timeline, the science behind the human genome project, advances based on human genome project, human genome project impact, the influence of human genome project on biological and technological progress, medicinal impact, effect on law and the sociologies, technical aspects of the human genome project, mapping strategies, sequencing strategies, findings of the human genome project, post-human genome project era: what is to come?

REFERENCES

Balmer, B. (1996). Managing mapping in the human genome project. *Soc. Stud. Sci., 26*(3), 531-573.
Cantor, C.R., Smith, C.L. (2004). *Genomics: the science and technology behind the human genome project.* John Wiley & Sons.
Collins, F. S. (1998). The human genome project: how private sector developments affect the government program. *Jun, 17,* 06-17.
Collins, F.S., McKusick, V.A. (2001). Implications of the human genome project for medical science. *JAMA, 285*(5), 540-544.
Collins, F.S., Patrinos, A., Jordan, E., Chakravarti, A., Gesteland, R., Walters, L. (1998). New goals for the US human genome project: 1998-2003. *Science, 282*(5389), 682-689.
Council, N.R. (1988). *Mapping and Sequencing the Human Genome..* National Academies Press.

Cowan, W.M., Kopnisky, K.L., Hyman, S.E. (2002). The human genome project and its impact on psychiatry. *Annu. Rev. Neurosci., 25*(1), 1-50.

Croce, N. (2015). *The Science and Technology Behind the Human Genome Project..* The Rosen Publishing Group, Inc..

Ellsworth, D.L., Hallman, D.M., Boerwinkle, E. (1997). Impact of the Human Genome Project on epidemiologic research. *Epidemiol. Rev., 19*(1), 3-13.

Gibbs, R.A. (2020). The human genome project changed everything. *Nat. Rev. Genet., 21*(10), 575-576.
http://dx.doi.org/10.1038/s41576-020-0275-3 PMID: 32770171

Ingelman-Sundberg, M. (2005). The human genome project and novel aspects of cytochrome P450 research. *Toxicol. Appl. Pharmacol., 207*(2) (Suppl.), 52-56.
http://dx.doi.org/10.1016/j.taap.2005.01.030 PMID: 15993453

Iscovich, J. (1998). Introduction to the post-Human Genome Project era, a target for interactions between polygenic and/or multiphenotypical components in cancer control in South America. *Cad. Saude Publica, 14* (Suppl. 3), 15-23.
http://dx.doi.org/10.1590/S0102-311X1998000700003 PMID: 9819461

Mourier, T., Hansen, A.J., Willerslev, E., Arctander, P. (2001). The Human Genome Project reveals a continuous transfer of large mitochondrial fragments to the nucleus. *Mol. Biol. Evol., 18*(9), 1833-1837.
http://dx.doi.org/10.1093/oxfordjournals.molbev.a003971 PMID: 11504863

Roberts, L., Davenport, R.J., Pennisi, E., Marshall, E. (2001). A history of the Human Genome Project. *Science, 291*(5507), 1195-1200.
http://dx.doi.org/10.1126/science.291.5507.1195 PMID: 11233436

Sawicki, M.P., Samara, G., Hurwitz, M., Passaro, E., Jr (1993). Human genome project. *Am. J. Surg., 165*(2), 258-264.
http://dx.doi.org/10.1016/S0002-9610(05)80522-7 PMID: 8427408

Kelavkar, U., Shah, K. (1998). Advances in the human genome project–A review. *Mol. Biol. Rep., 25*(1), 27-43.
http://dx.doi.org/10.1023/A:1006834711989 PMID: 9540065

Venter, J.C., Adams, M.D., Myers, E.W. (2001). The sequence of the human genome. *Science, 291*(5507), 1304-1351.

Waterston, R.H., Lander, E.S., Sulston, J.E. (2002). On the sequencing of the human genome. *Proc. Natl. Acad. Sci. USA, 99*(6), 3712-3716.
http://dx.doi.org/10.1073/pnas.042692499 PMID: 11880605

Williams, S.J., Hayward, N.K. (2001). The impact of the Human Genome Project on medical genetics. *Trends Mol. Med., 7*(5), 229-231.
http://dx.doi.org/10.1016/S1471-4914(01)02001-9 PMID: 11325635

Wilson, B.J., Nicholls, S.G. (2015). The Human Genome Project, and recent advances in personalized genomics. *Risk Manag. Healthc. Policy, 8*, 9-20.
http://dx.doi.org/10.2147/RMHP.S58728 PMID: 25733939

CHAPTER 9

Modern Genomic Implications

Abstract: This chapter proposes a brief description of Possible health enhancement; Genome-wide association studies (GWAS); GWAS and different case studies; Next-generation sequencing (NGS); Steps of Next-generation DNA sequencing; DNA library preparation, Amplification, Emulsion PCR, Bridge PCR, Sequencing; Impacts of Genomic in Future Health; Cancer: stratifying tumors for treatment; Drug prescription and development; Diagnosing and characterizing genetic disease; From disease diagnosis to personalized genetic health; Integrating genomic and clinical information; Economic benefits of genomic medicine; Targeted genome editing: a new era in modern genomics; Modifying the Genome Through Nucleases; Zinc Finger Nucleases, Transcription Activator-like Effector Nucleases, Clustered Regularly Interspaced Short Palindromic Repeats / Cas9 (CRISPR/Cas 9), Delivering the Nucleases; Application of Engineered Nucleases; Personalized Medicine; Epigenome.

Keywords: CRISPR/Cas 9, GWAS, NGS, Personalized Medicine.

The genomic advances are applied to the people that are living in developing countries for their health improvement. To guarantee that advantages are given to developing nations, consideration should be given to legitimate, social, and financial issues. Innovative and global mechanisms are expected to transform high expectations into the real world and permit the utilization of genomics to decrease well-being differences among rich and helpless countries.

The continuous revolution of genomics featured by sequencing the human genome will help change the diagnosis of diseases. It will improve health around the world. Now the medications from genomics will probably be costly, and these modernizations will mean the strength of individuals in the developing nations. Actually, a significant number of advances in genomics were made and, to a limited extent, are possessed by the modern world. The scientists are worried that the genomics divide will also broaden the value gap in well-being among rich and developing countries. According to the report released by WHO, it calls attention to that around eighty percent of the interests in genomics in 2000 was completed in the US, and 80% of DNA licenses in genomics in the time frame 1980 through 1993 were held by US companies. Two of the 1233 new drugs advertised somewhere

Maryam Javed, Asif Nadeem & Faiz-ul Hassan

between 1975 and 1999, just thirteen were permitted for the tropical infections (Zwart, 2009).

POSSIBLE HEALTH ENHANCEMENT

Almost 50 microbes have been sequenced. Now the efforts are being applied to genome sequence (Gs) of mosquito vectors, *e.g.*, Anopheles gambiae, the malarial vector, and Aedes aegypti, the fundamental vector for dengue fever guarantee aid in the control of infectious diseases. Fosmidomycin initially created with the aim at the therapy of urinary infections. It indicated a powerful enemy of malarial action when genome grouping data from P. falciparum uncovered a typical molecular target site present in the parasite and not in the human host. This drug has gone into clinical trials in less than 2 years. Likewise, clinical preliminaries have been ongoing in Africa for a pre-erythrocytic DNA-based antibody that gave critical defense against normal *P falciparum* infection. However, the advantages of reducing diseases are self-evident. It is currently accepted that data produced by genomics assume a significant part in the avoidance, analysis, and executives of numerous diseases that are difficult to control, including cardiovascular illness, malignant growth, significant psychoses, dementia, rheumatic illness, and asthma. From a general health viewpoint, the genomics revolt introduces new chances for the anticipation of these diseases. However, before these changes can be acknowledged, we should determine what grouping of hereditary and environmental elements incline individuals to such diseases. Another grouping has been shaped to seek after this methodology which plans to rule the significant innovations to test sizes fitting for epidemiologic studies. However, the underlying spotlight on diabetes and cardiovascular disease are to create tools that eventually apply to different diseases. Three billion individuals basically rely upon rice as their staple eating regimen. The rice sequencing genome may prepare for improved strains of rice that give improved crops, healthy and beneficial.

Besides the complex logical and practical issues of carrying genomics to clinics, guaranteeing the advantages of developing nations will expect consideration regarding numerous corresponding testing issues. Genomics carries with its complex new moral, legitimate, social, and financial complications, just as worries about dangers and hazards. There are cases of secrecy, abuse, and mishandling of hereditary data, especially to create a genomic class that might be prevented by clinical protection from getting genetic testing and screening. Genomics has also been related to the possibility of originator infants. There is a corresponding worry about making a hereditarily designed overclass and a sickness inclined underclass; the higher probability of the previous being related with more extravagant individuals in the developed world is self-evident. Issues of licensed innovation rights related with DNA sequences and the expected abuse of developing-nation by making hereditary information bases, frequently at the command of organizations in the developed world. While industry accepts that without solid and viable worldwide licensed innovation runs, the hole among developed and non-developed nations will fill later. There are many worries about the DNA patentability sequencing and applications got from them and what suggestions this will have for the non-developed nations. Generally, the moderately rich element pipeline of genomics-based medications will mean a gigantic expansion in the interest for clinical destinations. Large numbers of which will be in developing nations. This region addresses a moral minefield identifying issues, such as up-to-date consent, standard of care, and accessibility of medication tried—the cost of which regularly past the compass of poor people. Lastly, in the outcome of the misfortune that occurred in the US on Sep. 11, 2001, the usage of developments in genomics for demonstrations of biological terrorism and organic fighting comparatively consumes the minds of many.

Despite the gigantic development and guarantee of genomics, it is exceptionally hard to anticipate its advantages for health There are such countless crucial things we don't until now think about how products of gene associate. Numerous individuals were shocked to discover that we have just twice, however, many genes as a fly or a worm. Henceforth it is extremely significant for the agricultural nations to keep up the spotlight on the essentials of what should be possible now, especially in the fields of general health and the advancement of more practical medical services frameworks. The fundamental memo of the WHO is that clinical practice won't change for the time being because of new advancements generated by genomics, yet the drawn-out potential outcomes are to such an extent that both nations should set themselves for this innovation and cautiously investigate its prospects, continually seeing its expense adequacy contrasted with more standard methodologies and clinical consideration. Likewise, it is essential that genomics research not be sought after to hinder the grounded strategies for clinical practice and epidemiological study. Surely, it should be coordinated into clinical examination for its full misuse, including patients and epidemiological investigations locally. It is essential to keep an equilibrium in clinical practice and exploring genomics and these more ordinary and well-adopted methodologies. Furthermore, it is critical to building the Nature of instruction in genomics. If this isn't accomplished, it will be difficult to build up a knowledgeable discussion about different issues included and there's a hazard that the individuals who manage health administrations will not be able to recognize metaphor dubious and quickly growing research field (Bush & Moore, 2012).

GENOME-WIDE ASSOCIATION STUDIES (GWAS)

Genome-wide association (GWAS) is a new approach in genetics research to associate distinct genetic variation with a specific disease. GWAS involves the fast scanning of up to 5 million markers across the DNA in many people that can predict the presence of various diseases. By GWAS, several genetic risk factors can be found for common diseases and are widely used to identify genes responsible for human diseases. This method searches the genome for small genetic variations known as single nucleotide polymorphism (SNPs). These SNPs are found more frequently in affected individuals than normal individuals. The genetic variants identification was made possible by rapid massive parallel genotyping and, international HapMap project. The HapMap project was a huge international collaboration that started in 2002s, and its basic objective was to search for common genetic variant patterns of DNA in the human genome. It has been done by characterizing various genetic variants, frequencies, and associations in DNA samples from a population with lineage from Africa, Europe, and Asia. As the human genome project was a huge achievement, this project was also a victorious collaboration. This genetic variation directory has contributed to the requisite 'blueprints' for genetic variation, making the GWAS possible. After a year, the 2000s genotyping technologies have advanced and developed, permitting genome investigation that was impossible in the past. Nowadays, many commercial companies can provide high-density SNP genotyping from 20,000 to 1.2 million SNPs (Tam *et al.*, 2019). GWAS study design consists of the following steps.

- The selection of the appropriate number of individuals (for instance, cases, control for disease, and a sample of the unselected population for a trait) for the appropriate comparison group in a binary situation.

- Isolation of DNA, genotyping (by SNP arrays or WGS)

- Quality control analysis.

- Statistical tests for association between SNPs.

- Replication of known association in the independent population sample.

Generally, GWAS uses a case-control study design with a huge number of unrelated members and then compares the frequency of single nucleotide polymorphism in a defined group of individuals with those whose phenotype is in question. If a specific pattern of genetic variation is more common among affected individuals, it will be the causative agent or linked with the causative mutation. By appropriate selection of the population, genetic variants and biological pathways of a particular disease can be determined. By this approach, several SNPs have been identified associated with complex diseases such as diabetes, Parkinson's disease, heart disease, and many other chronic diseases (Ku, Loy, Pawitan, & Chia, 2010).

GWAS and Different Case Studies

Asthma is a severe disease of an airway with heterogeneous phenotypes, and multiple genes are responsible for this chronic disease. As it is studied, the GWAS uses high-throughput sequencing technology. We use this to identify the genes related to asthma susceptibility. It has become possible to find asthma-associated loci and specific genes correlated with pathogenesis. One of the GWAS previous studies shows that five loci are responsible for asthma exacerbation. These loci are IL33, RAD50, GSDMB, CDHR3, and IL1RL1 linked with asthma susceptibility. In another GWAS study, CTNNA3 and SEMA3D, these two genetic variants were reported as risk loci. Similarly, in adults, PTTG1IP and MAML3 and PTTG1IP loci were reported associated with bronchial hyper responsiveness (BHR).

Breast cancer is the leading cause of death in the world. The first genome-wide association study for breast cancer was performed in 2007, and novel loci associated with the disease were identified. It is reported that the genes that are present around 4 loci are the reasonable causative agent (TNRC9 FGFR2, LSP1, and MP3K1). The most powerful SNP is associated with the introns of the FGFR2 gene and tyrosine receptor kinase that are overexpressed in 5-10% of breast cancer. It is reported that more than 170 genomic loci consisting of common variants have been identified in breast cancer, of which twenty are primarily linked with estrogen receptor (ER) negative disease (Stadler *et al.*, 2010).

It is reported that GWAS has been widely used in investigating the pathogenesis of ophthalmic conditions. One of the GWAS was conducted to find a refractive error in the European population. It was reported that many polymorphisms at 15q25 near the *RASGRF1* gene were linked with ocular refraction. In the human retina, *RASGRF1* gene is highly expressed.

Agriculture production is affected by various biotic and abiotic stresses. GWAS is used as a robust tool for identifying complex or single traits related to any type of stress. GWAS on various plants identified the unique gene candidates that are responsible for biotic and abiotic stress. The result of these investigations will be beneficial for the breeders. It can be used to investigate drought tolerance, salt tolerance, thermal tolerance, and other agronomic and flowering traits. It was reported by reported in previous studies that in Arabidopsis, thaliana proline is responsible for the drought response due to low water potential. GWAS can identify salt-tolerant genes at the seedling stage. Similarly, another study reported that 25 quantitative trait loci based on 75 Single nucleotide polymorphisms spread over 14 chromosomes, .and these were associated with 4 salt-tolerant genes.

Although some limitations are associated with GWAS, such as Ultra-rare mutation cannot be identified by this and is not successful in detecting epistasis in humans. Similarly, GWAS identifies whole genetic determinants of complex traits, and Population stratification is the major limitation of GWAS. With the emergence of Genome-wide association studies, it has become possible to detect the genetic basis of disease. However, some limitations are associated with it. Still, it can be overcome in the future by advanced technologies as the sequencing techniques are getting faster and cheaper; we will be able to understand human genetics with more accuracy. In the future, we will be able to design more therapeutics and personalized medication for many lethal diseases with these high-throughput technologies. As we detect the stress-related genes in plants, it will be beneficial for advanced breeding and designing future crops with disease-resistant traits (Scherer & Visscher, 2016).

NEXT-GENERATION SEQUENCING (NGS)

First-generation sequencing includes Sanger and Maxam Gilbert Sequencing methods. Both methods were accepted earlier, but Sanger sequencing, also known as the chain termination method, is used for a routine sequencing basis. The maximum read length generated by these methods is 800-1000bp and is quite time-consuming. Therefore, first-generation sequencing is replaced by Next-generation sequencing technology. NGS is a comprehensive DNA technology that makes it possible for querying the whole genome, whole exome (exons in whole known genes), or only exons of a targeted genome. Using NGS, we can interpret millions of genes and sequences of nucleotides simultaneously, and therefore this technique of sequencing is known as massively parallel sequencing. In ancient times before the 2000s, the complete sequencing of the human genome gave rise to a need for a high-throughput and cheap sequencing method because traditional sequencing was quite expensive and time-consuming. This was overcome by NGS platform and was rapidly commercialized. This technique is briefly described as First, from a patient sample, DNA library is prepared by disintegration, purification, and then amplifying a sample. Each fragment is physically separated on small beads or solid surfaces. In this way sequence of every single fragment can be determined as sequencing by synthesis then obtained sequence data is computationally aligned against a normal reference human genome. By this technique, in a single reaction, various sequence alterations can be detected (Behjati & Tarpey, 2013).

Steps of Next-generation DNA Sequencing

- DNA library preparation
- Amplification
- Sequencing

DNA Library Preparation

First, fragmentation of DNA is carried out either by enzymes or by sonication to smaller fragments. Then with the help of DNA ligase adaptors, smaller pieces of synthetic DNA are ligated to these fragments of DNA. These adaptors can be bound to their complementary parts, and ligation occurs between 3-OH and 5-P ends. For sequencing purposes, these libraries are clustered in PCR colonies. The library construction by this method is faster and cheaper than the traditional sequencing method; moreover, there is an expense of fragment read length.

Amplification

Amplification of library is needed so that signal received from the sequencer is adequate to be detected precisely. Amplification *via* enzymatic methods biasing and duplication phenomenon can occur, leading to preferential amplification of some DNA fragments. Alternatively, various types of amplification methods use PCR to create a huge number of clusters. The PCR used are:

- Emulsion PCR
- Bridge PCR

Emulsion PCR

Emulsion oil, PCR mix, beads, and library DNA are mixed to make an emulsion, leading to microwell formation. For the sequencing process, every microwell should consist of one bead with one strand of DNA. The library fragments are then denatured in PCR and behave as two isolated strands and annealing of one reverse strand to the beads. By polymerase, this DNA is amplified initiating from beads towards the primer site. Then actual reverse strands are denatured and free from the beads. This process is repeated 30-60 cycles which leads to the huge collection of DNA. However, this technique requires a lot of time in bead formation, breaking, and amplification despite this being used in many next-generation sequencing platforms (Gansauge *et al.*, 2017).

Bridge PCR

The cell surface is tightly coated with the primers which are complementary to the primer attached with DNA fragments. Then randomly, this DNA attaches to the cell surface for the polymerase-based extension. Furthermore, in addition to enzymes and nucleotides, the free ends of single-stranded DNA attach themselves to the cell surface to the primer, which is complementary to them by forming bridge structures. Then bridges are converted into double-stranded DNA by enzymes, and when denaturation occurs, these two single strands on the cell's surface are close to each other. By repeating this process leads to clusters of identical DNA (Shao *et al.*, 2011); (Mehta & Singh, 1999).

Sequencing

There are different methods or strategies are used for sequencing by different companies. These methods are given as:

- 454 Pyrosequencing
- Ion Torrent semiconductor sequencing
- Sequencing by ligation (SOLiD)
- Reversible terminator sequencing (Illumina)

Pyrosequencing is based on the principle known as sequencing by synthesis, where the enzyme forms the complementary strand. However, as in sanger sequencing, chain termination by dideoxy nucleotides terminate the amplification of the chain, but in pyrosequencing, a pyrophosphate is released when nucleotide addition takes place at the DNA chain. Initially, it uses an emulsion PCR process to construct PCR colonies or polonies required for sequencing. After this, as the single-stranded DNA primer binds to the primer binding region, then four various dNTPs are continuously made to flow in and out of the wells. When the correct dNTP is added, pyrophosphate is released. Then pyrophosphate is transformed into ATP in the presence of adenosine and ATP sulfurylase. This ATP molecule is utilized to convert luciferin into oxyluciferin, which turns out the camera can determine the light. As more bases are added, the relative light intensity will be increased.

Ion Torrent, semiconductor sequencing technology, is also based on sequencing by synthesis principle. It uses the semiconductor chip, which detects the production of hydrogen ions that are produced during DNA polymerization. In this approach, as the new dNTP is added, the hydrogen ion is released. More dNTPs added more hydrogen ions would be produced, which can be detected by the pH sensor in the sequencer. Although this is a faster and cheaper sequencing method, it is difficult to enumerate identical bases that are added consecutively.

Sequencing by ligation (SOLiD) is an enzymatic method of sequencing that utilizes DNA ligase. PCR emulsion is used to amplify single-stranded DNA primer binding regions known as adopter sequences, attaching to the target sequences on the beads. These beads are then placed onto the glass surface, and beads with high density can be achieved, enhancing the high throughput of this method. This sequencing reaction is divided into 5 further steps: preparation of DNA library, amplification by emulsion PCR, attachment of beads, sequencing, and then resetting of primers (Mattick *et al.*, 2014).

NGS has many applications in different fields, such as being used to characterize food-related biomes for the detection of sequence mutations which is commonly used for disease diagnosis, prognosis, and therapeutic decisions. Similarly, also have application for surveillance of environmental spread of antimicrobial resistance genes, for the detection of unknown viral pathogens, uncovering of Novel Viruses, and monitoring antiviral drug resistance, with the advent of next-generation sequencing technologies. It has become easier to find unknown species and many other complex diseases. The different companies use different platforms depending on the read length and number of the reads to verify the quality of assembly and accuracy. So, for future research, it is necessary to improve methods for analysis for the huge amount of data created by NGS. In the main future objective will be to reduce the time of processing, increase the precision of sequencing of assembly and optimize algorithm efficiency for analysis. Eventually, human genomic information can be revealed by NGS, and functions of the human genome can be elucidated, which may provide a better approach in personalized medicine (Mardis, 2008).

IMPACTS OF GENOMIC IN FUTURE HEALTH

Genomic medicine will groundbreakingly mark individual health and prosperity. Lately, there has been a remarkable jump in evidence on the human genome and its part in health. Ten years prior, experts were likely investigating the first reference genome of human sequencing, which cost more than one billion dollars to produce. Now many genomes from ethnic families have been sequenced. This blast action has been empowered by remarkable advances in sequencing that will be able to sequence an individual's whole genome over six thousand million bases in days at the expense of one thousand and three US dollars. Figuring out genomic information requires computational innovations and data sets to develop corresponding with sequencing skills. Advances in the two skills empower an increasing limit regarding the precise finding of current disease and improvement of powerful and aim at treatment methods. They likewise suggest surveying an infection, possibly provoking more engaged clinical nursing and way of life changes. Despite the fact that our struggle into the human genome is a long way from complete, models show that even our restricted genomic comprehension can be amazing in the facility. Right now, (GS) has good results in delineating cancer growth, describing genetic infection, and giving data about a person's possible reaction to treatment (McBride *et al.*, 2010).

Cancer: Stratifying Tumors for Treatment

Genomic medication has just demonstrated an advantage in analyzing and managing remedial methodologies for Cancer. Since the last part of the 1990s, the clinician's malignancy "toolkit " of medical procedure, radiation, and chemotherapy has been progressively enhanced by treatments that target exact molecular markers of cancer growth. Information of genome would now help clinicians choose therapy techniques by characterizing a tumor as per its changes and relating drug sensitivities. Sometimes, patients have been saved with expensive and complex methodology, such as bone marrow transfers, in light of a molecular diagnosis. In different models, patient malignant growth advancement has been balanced for a period by focusing on target particles in the tumor cells. Clinicians are beginning to utilize genomic data to improve disease diagnosis and modify a person's personal life. Genome-wide sequencing is currently being applied to investigate circling DNA in the plasma of disease patients, just as in people with other diseases. This innovation empowers non-intrusive tumor location and checks reactions to treatment that guarantees to enhance patient life essentially (Yeo & Guan, 2017).

Drug Prescription and Development

Pharmacogenomic applications reach out into numerous parts of clinical practice. In the future, genomic data is relied upon to significantly change the testing and utilization of drugs through infection stratification. In corresponding, scientists are utilizing genomic information to propose new helpful applications for existing medications with tremendous expensive investment and to choose people for clinical preliminaries to discover more readily utilizes for drugs that were failed before. Better focusing on existing medication use maintains a strategic distance from inefficient and unsafe treatment and will be cost-effective (Kessel & Frank, 2007).

Diagnosing and Characterizing Genetic Disease

In Australia, monogenic and hereditary diseases are tested. Each infant in Australia is screened for around 30 hereditary conditions in the Guthrie test, and more than 300 tests for hereditary problems are accessible through the medical services system. They are utilizing this data to direct treatment. As the field develops and our comprehension of the influenced pathways improves, more genomic loci will be involved in uncommon diseases, with improved therapy possibilities. Further, genome sequencing is progressively being utilized to survey hereditary complex diseases, where numerous gene variations might be associated with infection advancement. Instead of sequencing the whole genome, sequencing of the protein-coding segment (the exome, about 1.5% of the aggregate) can offer a relatively financially savvy examination of numerous hereditary problems. Although less viable in identifying chromosomal duplications and rearrangements, in a single analysis, exome sequencing recognizes the most clinically important nucleotide variations serving both as a symptomatic device and a technique to find new mutations and genes. In any case, as the capacity of intergenic and intronic areas in infection- regulatory cycles is additionally explained, and variations in these areas are appeared to have diagnostic use, entire genome sequencing is probably going to give a clinically more hearty, exact, and supportive test that will substitute exome sequencing as sequencing costs decline.

From Disease Diagnosis to Personalized Genetic Health

It is getting more normal for people to look for individual genomic data through genetic testing services (GTS). In any case, these tests are not certified for the clinical conclusion. People will acquire the most advantage when their genomic clinicians and other health experts can give thought about counsel and the gateway to treatment alternatives. An individual genome sequence is a very important part of an electronic health record (EHR) that will be coordinated with other clinical and ecological information and examined throughout the person's life. Clinicians and patients may precisely recommend medicines, decide infection susceptibilities and recognize drug sensitivities, and decide a strategy to screen, oversee, enhance the danger of or prevent the illness. Sequencing newborn babies stay disputable. There are additionally viable inquiries regarding information storage and access and the subject of which results are returned when and to whom (Mattick *et al.*, 2014).

Integrating Genomic and Clinical Information

The lane from the DNA sequencing of the patient to the plan of clinical treatment is based on the integration of known databases of the individual's genome comprised of correlations of genotype and phenotype that are known and the individual's study of association from a broad group of population. Clinical dynamic as of now depends on the information provided by the individual experts and hereditary guides. These experts function as a multidisciplinary group to evaluate the variations in genomics recognized through genome sequencing and show up at a treatment plan. There is a need for reason constructed, well-curated, and persistently refreshed proof-based data sets of genotype and phenotype associations studies of humans in combination with tools of bioinformatics to cross-examine the steadily expanding data in a mechanized manner. The inherent challenges in computational studies in coordinating genomic and clinical information require a huge investment in bioinformatics ability. Australia will most successfully be served as a repository of unified genome information connected to worldwide vaults if provided with significant overheads of both powers of bioinformatics and storage. Advances in E-Systems and programming interfaces highly need the support of infrastructure development. These should empower clinicians and medical services suppliers to investigate the sequences of the genome of patients against the clinical choice help data sets to get an instructive clinical report.

Economic Benefits of Genomic Medicine

Genomic medication will change medical services and the public economy, particularly in a populace whose normal life expectancy is expanding. Individual financial advantages accumulate from gnomically educated reclamation regarding well-being and subsequent acquiring limit. Health cost reduces by higher precision in identifying disease risk of a person and in the system of healthcare by preventing pointless therapies and adverse reactions. Genomic medication can possibly make the hereditary determination of illness a more productive and savvy measure by diminishing hereditary testing to a solitary investigation, which at that point advises people all through life. Despite the fact that people will differ in their reaction to genomic data, individual recognizable risk could be required to take up more viable observing and preventive activities. Applications of the genome for technical developments, clinical examination, and medical services will likewise significantly affect the nation's economy by lessening productivity misfortunes, diminishing expenses of treating sickness, and making new clinical data ventures. The quick expansion in population's age, quickening expenses of medical care, and the increasing weight of chronic diseases depict significant difficulties in the well-being frameworks around the world. The UK government has perceived this goal, as of late reporting its interest in a task, to be controlled by Genomics England, to sequence the genome of patients nearly 100000 more than 5 years and bring genomic innovation into its standard well-being framework (Feero, Wicklund, & Veenstra, 2013).

TARGETED GENOME EDITING: A NEW ERA IN MODERN GENOMICS

The CRISPR-CAS system has revolutionized genome editing. This system has been extensively used in the treatment of different diseases. , The development of new efficient and reliable approaches is a long-standing goal for biomedical researchers that are proficient in bringing specific changes in targeted sequences in the genome of living cells. This can be achieved by using engineered nucleases (Biological scissors) to produce precise double-stranded breaks on the targeted site of the genome and coupling it with the cells repair mechanism of homo-genius recombination and non-homologous ends joining. Currently, four families of programmable nuclease like transcription activator-like effector nucleases (TALENs), Clustered Regularly Interspaced Short Palindromic Repeats/Cas9 (CRISPR/Cas9), zinc-finger nucleases (ZFNs) and meganucleases have been used in genome editing. The significance of these modular assemblies in medicinal and biotechnological research stems from potential end routes for promoting sites directed DNA cleavage in the genome, and consequential repairing (*via* endogenous mechanisms) permits meticulous genetic modification. Yet, these nucleases vary in numerous aspects, including their composition, possible target sites, target sites length (specificity determinant), recognition sites and mutation signatures, along with other features. Understanding the specific properties of nuclease and their advantages and disadvantages is vital among scientists to decide the most suitable tools for a diverse variety of implementations.

The foremost endeavor of genetics is the association of genotype and phenotype. The conventional strategy is to identify a unique phenotype, whether naturally or conceived by mutagenesis, to classify the responsible gene and discern why mutations at the site have the experiential effect. In relation, the contemporary approach often referred to as reverse genetics involves selecting a gene from the genomic sequences to be introduced with site-specific mutations and the characterization of the consequential phenotype. Defined alteration of sites inside a gene of interest is a typical approach to reveal the gene function, construct disease animal models, and improve desired features of animals and plants. Targeted genome engineering (i.e., genome alteration at a specific, predestined locus) is largely pertinent to medicine, biomedical research, and biotechnology. Previous approaches for defined genome modifications *via* embryonic stem cells by homologous recombination were of limited applicability, with the only successful results in mouse ESCs. It has been demonstrated that introducing double-strand breaks at a precise site in the genome could excite the cellular DNA repair mechanisms and boost the rate of homologous recombination at the break point. Vitally, this study also proposed a direct approach for modifying practically any genome by means of homologous recombination if the DNA breaks could be directed to precise sequences. The re-designing of molecular apparatus to distinguish novel sequences in intricate genomes is a tedious task. Nevertheless, the discovery of the crystal structure of Zif268 in 1991 (a naturally occurring zinc finger protein) gave insight into how Nature resolved this dilemma (Ghosh, Kumar, & Sinha, 2021).

Modifying the Genome Through Nucleases

Over expression" and "loss of function" are the two key routes generally employed to explore the function of a gene. The first approach involves the forced expression of the gene of interest in targeted cells or transgenic animals, while the second one is concerned with the loss of activity of gene to develop a model that would serve to study human genetic disease caused by mutations. Both approaches facilitate the analysis of how genotypes may affect the phenotype. The former can be achieved by direct gene transfer using suitable vectors. Likewise, transgenic animals could be generated through stable germ-line integration of the transgene. The latter approach could only be implemented by targeted mutations in the genome using "homologous recombination" (HR). This method allows the introduction of null mutations as well as targeted sequence alterations in the desired gene. Essentially, HR is the strand switch over between identical DNA sequences. Thus, a synthetic DNA molecule (donor vector) with identical sequences may cause HR in the specific locus with sequence similarity. However, this is an exceptional event under a physiological environment. Even when well-identical sequences are present, mammalian genomic DNA is very stable, and natural recombination virtually never takes place. In the physiological condition, the rate of recurrence of homologous recombination is expected to be one in every 10^4 to 10^7 cells. Nevertheless, mammalian cells may use homologous recombination in DNA repair mechanisms. The efficacy of homologous recombination can be enhanced by pursuing the non-physiological pathway. The breaks in the double-strand boost the likelihood of repair by homologous recombination [23]. The modern applications developed in the past fifteen years, termed as "genome editing tools," allow the attainment of targeted modification in any gene, both *in vitro* and *in vivo*. Through these approaches, defined double-strand DNA breaks can be introduced in a target sequence. This is achieved by using the "nuclease" function that results in the hydrolysis of the phosphodiester bonds on both strands of a DNA helix.

As mentioned earlier, such double-strand breaks induce DNA damage response triggering the HR in the process. The crucial step here is to direct this nuclease action towards the target DNA sequence. The nucleases engineered based on enhanced DNA recognition property regulate sequence specificity. This way, proficient and precise genetic modifications can be attained at a definite locus. Initially, DSBs are introduced at the target locus, followed by the induction of cellular DNA repair mechanisms to repair. The type of the genome-editing technique depends on the preference of the repair process. Either a more fallible "non-homologous end-joining" (NHEJ) or HR is initiated. The earlier NHEJ, swiftly re-ligates the broken ends. An imprecise repair mechanism results in deletions or insertions of one or more bases at the repair site (Carroll, 2014).

Suppose these sequence modifications take place within the coding regions of the target gene. In that case, they will function as frame shift mutations producing a truncated, apparently non-functional gene product (loss of function mutation). The process can be exploited for determining knockout phenotypes. On the other hand, when a donor DNA sequence is present for homologous recombination, homology-directed repair (HDR) is then activated. Any required sequence alteration can be achieved by this engineered HDR approach. The application of the targeted nucleases with or without recombination vectors is called "genome editing". Currently, genome editing tools may be utilized to produce targeted mutations, gene corrections, and gene substitution. To achieve sequence specificity to the nuclease action, site-specific DNA-binding proteins are employed. These proteins are derived from

- zinc-finger proteins (ZFPs)
- transcription activator-like effectors (TALE) protein
- clustered regulatory interspaced short palindromic repeats (CRISPR/Cas). These novel technologies are efficiently used in various cell types and organisms. These novel tools expedite the genetic engineering studies harboring the promising clinical potential in gene replacement therapy.

Zinc Finger Nucleases

The classical Cys2His2 zinc finger is possible by elucidating structural and biochemical examination of the structure of *Xenopus* protein transcription factor IIIA and its interactions with 5S RNA and DNA. The fingers are self-regulating domains stabilized *via* the binding of zinc ions to two cysteines and histidine. This discovery led to the finding of new protein folds and a new principle for DNA recognition. While various DNA binding proteins usually rely on the two-fold symmetry of the double helix, zinc fingers may join sequentially in tandem to identify various sequences in nucleic. Such modularity in designs holds the potential for generating greater combinatorial options for the specific recognition of DNA (or RNA). Therefore, the first engineered zinc-finger nucleases (ZFNs) were based on the DNA sequence recognition property of the zinc-finger transcription factors. Zinc-finger proteins, as stated above, have a sequence-specific

DNA-binding activity like other transcription factors. In 1994, a study was carried out to generate a chimeric protein by hybridizing the zinc-finger domain with a nuclease to create sequence-specific DSBs. Their first application to a biological system resulted in constructing a three-finger protein that aimed to halt an oncogene's expression in transformed mouse cell lines. The two independent domains, namely the DNA-binding and the DNA cleavage domains, comprise the fundamental ZFNs structure. The cleavage domain is principally a nuclease that lacks any selective reactivity for the target sequence. The DNA-binding of the zinc-finger domain projects cleavage domain to the specific locus (Carbery *et al.*, 2010).

Transcription Activator-like Effector Nucleases

Zinc-finger nucleases were developed as genome editing tools 19 years ago and since then, they have been used to manipulate yeast, C elegans, zebrafish, drosophila, and plant genomes. However, their widespread use has been partially hindered by their time-consuming, complex, and expensive Nature. Several studies carried out in the previous ten years have established the immense potential of custom nucleases in genome editing. TALEs can prove to be a useful alternative to ZFNs. TALENs have a similar structural organization to that of ZFNs as they also possess FokI nuclease domain. However, the DNA binding domains, transcription activator-like effectors (TALEs), the derivative of a plant's pathogen Xanthomonas bacteria, differ in TALENs. TALE-like proteins from other phytopathogenic bacteria *Ralstonia* can also be developed to recognize specific DNA sequences. TALENs have been employed to introduce modification in the genes of several species such as viruses, yeast, nematodes, frogs, plants, insects, fish, and mammals such as rats, mice, pigs even cultured mammalian cells. TALEs contain 33–35 amino acid repeats arranged in a tandem manner, each one interacting with a single base pair in major grooves of the DNA. These repeats are almost similar in sequence except for the amino acid residues at 12th and 13th positions, which establish the nucleotide specificity of every repeating domain. These two-hyper variable amino acids are called repeat-variable residues (RVDs). Asn-Asn, Asn-Ile, His-Asp, and Asn-Gly have extensively used specific nucleotide binding RVD modules and correspond to Guanine, Adenine, Cytosine, and Thymine, respectively (Sun & Zhao, 2013).

Clustered Regularly Interspaced Short Palindromic Repeats/Cas9 (CRISPR/ Cas 9)

The latest entry to the programmable nucleases group is derivative of the adaptive immune system used by the bacteria and archaea against invading DNA or RNA segments of viruses or plasmids. These microorganisms incorporate short foreign DNA sequences (20–50 bp) into their personal genomes between certain repeat sequences (also 20–50 bp), forming arrays that are called CRISPRs (clustered regularly interspaced short palindromic repeats). Upon subsequent invasion by a similar virus or plasmid, the related spacer sequences guide the cleavage of the foreign genome, thus fighting the invasion. The elements required by the CRISPR Type-II system for cleavage are namely CRISPR RNAs (crRNA), transactivating CRISPR RNA (tracer) and Cas9 nuclease. The processed RNA transcripts of the CRISPR array (crRNA) harbors a variable segment transcribed from the foreign DNA (protospacer sequence) and the portion of the CRISPR repeat. The crRNA hybridizes with another partial complementary RNA sequence called a tracer, to form a duplex. The duplex complexes with protein Cas9 contain a double nuclease active site, cleaving one strand of target DNA. Cas9 is primarily directed through this crRNA/ tracrRNA hybrid and needs a small sequence called proto-spacer adjacent motif (PAM) in target directly downstream the hybrid. The crRNA & tracrRNA of *Streptococcus pyogenes* has been made in combination with single-guide RNA (sgRNA) that provoke proficient Cas9 cleavage *in vitro*. The -editing potential of CRISPR/Cas system was immediately realized and applied for genomic manipulation of humans, zebrafish, and many other genetic make-ups. In all these applications, sgRNA has twenty directing sequences complementary to a specific target followed by 80 nt of crRNA/ tracrRNA duplex. This sgRNA construct is frequently cited as "+85" in previous studies (Naeimi Kararoudi *et al.*, 2018).

The fundamental strategy is relatively simple. Based on the researcher's objectives, a target position is chosen. It is essential that apart from complementary 20 bp directed to sgRNA, PAM should be located appropriately. The sequences for *S. pyogenes* Cas9 in this regard is (5' NGG). In some studies, DNA/RNA hybrid of less than twenty base-pair has been implemented very successfully sgRNAs can be simply created *via* transcription through both *in vitro* and *in vivo*, a promoter which is employed will define the base(s) at the 5'end (1 G for RNA polymerase III promoter, 2 Gs for the commonly used T3, SP6 & T7). At first, scientists were cautious about following this base to go with the target, but this proved to be unnecessary later on. Initial analysis of the Cas9 protein crystal structures, alone, sgRNA-complex, and a single strand of target DNA, and cryo-electron microscopy reconstructions revealed the protein domains arrangement and its recognition and cleavage activity.

Due to the versatile Nature of protein's activities, only a small portion of the protein sequence is essential. A large portion of the Cas9 binds with the RNA/DNA hybrid in addition to forming brief contacts with the invariable segment of the sgRNA and isolated domain for cleavages of 2 target strands. Such configurations present opportunities for developing a version of Cas9 that can provide altered PAM recognition and improved specificity amongst additional important features (Detsika *et al.*, 2021). Major advantages of CRISPR/Cas system over the ZFNs and TALENs are listed below.

1. Single invariable proteins are needed, thus eliminating the requirement for protein engineering.
2. Targeting depends upon complementary base pairing, so sgRNA development needs the solitary understanding for Watson–Crick rules.
3. Novel sgRNAs can be simply created.
4. For the above-mentioned advantages, it is practicable to outbreak several targets concurrently using diverse sgRNAs.

Delivering the Nucleases

ZFNs, TALENs, and CRISPR/Cas systems are employed in targeted genome engineering in different organisms. In every biological system (cell or organism), the successful genome alteration relied on the ability to transport all the reagents proficiently while conserving the functionality. The nucleases could be delivered and expressed *via* viral vector delivery, plasmid-DNA transfection, and transfection with synthetic mRNA in the case of cell cultures. The nuclease coding sequences must be delivered along with suitable expression apparatus such as promoter and polyA addition signals. The proteins must hold a nuclear localization signal. The ZFN protein naturally harbors positive surplus charges that facilitate their uptake by the cells readily from the culture medium, but it's not so in the case of TALENs or Cas9. Large donor DNAs could be delivered through plasmids or vectors, but oligonucleotide donors can only be supplied alone in the medium (Hu, Davis, & Liu, 2016).

When targeting the whole genome of the organisms *via* nucleases, various methods are essential for proper delivery. But all these methods should amend according to the suitability of specific conditions. Sometimes, direct delivery of nuclease mRNAs or DNA expression constructs into embryos through injection is very successful. For example, mRNAs are efficiently injected into zebrafish embryos just after fertilization. Protein expression at early stages directs both the somatic mutation and germ-line mutations through NHEJ. The same method has been efficiently applied to numerous organisms, including mammals. *Drosophila* embryos are injected at the multinucleate stage. Recovery of transmissible mutations was accelerated by injecting from the posterior side of the precursor (germline). The donor DNAs can be generally mixed into the mRNA injection mix for all systems. Under various circumstances, the organism's biology impedes the direct approach to delivery. Despite initial successful NHEJ mutagenesis *via* ZFNs in zebrafish, researchers could not attain the mending from the donor DNA template. After introducing an oligonucleotide donor after TALEN, Cas9 cleavage, numerous products simply showed partial corresponding matching; they were homologous on one end but non-homologous on the other. Completely homologous products have been acquired using a lengthy donor molecule, even though at lower levels than NHEJ mutants. Embryos of the Zebrafish undergo rapid cell divisions after fertilization. Within this stage, rapid repairing of DSBs through NHEJ appears to be suitable to more conscious, template-guided HR (Chardon *et al.*, 2019).

A similar thought-provoking situation is seen in common nematode *Caenorhabditis elegans*. Somatic mutagenesis was efficiently attained with ZFNs, at genomic and extrachromosomal targets. Such worms repress the expression of the transgene in the germ through RNA interference, which probably perturbed initial trials to yield transmissible mutations. In recent times, mRNA or DNA injections are significantly used to accomplish germ-line mutagenesis, HR, with all other nuclease platforms used innovative scheme, injection of Cas9 protein-sgRNA-complexes, without any requirement of mRNA translation hence evasion of RNAi. Unexpectedly, in some instances, mutagenesis happened in the germline of the F1, signifying the persistence of nuclease persists for several days. Nonetheless, on average, not higher, the recurrences being effectively modified worms can be isolated *via* screening.

Application of Engineered Nucleases

Zinc-finger nucleases (ZFNs) are nowadays rarely used in gene cloning as a substitute for restriction enzymes. Various genomes such as mosquitos, crickets, silkworms, pigs, cows, rabbits, and non-human primates have been altered using these programmed nucleases. Engineered nucleases are employed for manipulation and disruption in protein-coding genes and non-coding elements like micro RNAs. The physiological functions of single-nucleotide polymorphisms can be explored by introducing these genetic variations in various cell lines and model organisms *via* engineered nucleases. Gene knocks out or knocks in models can be created by directly injecting the engineered nucleases in single-cell stage embryos evading the need for embryonic stem cells. Thus, the programmed nucleases prove to be an indispensable tool for developing disease models in human pluripotent stem cells and numerous animals. The animal models may be applicable in preclinical drug tests drug target validation. Moreover, when introduced in human pluripotent stem cells, *in vitro* human disease models with isogenic controls, also referred to as "patients in a test tube" can be created, which aids the examination of disease physiopathology and drug testing. Pathogen-resistant rice has been developed by distorting pathogen-associated DNA sequences in rice genome by applying transcription activator-like effector nucleases. ZFNs enabled the removal of glycol antigens in pigs, triggering acute immune rejection in humans, making pig to human xenotransplantation more feasible. Finally, cell lines for therapeutic protein production, such as Chinese hamster ovary cells, have been established due to genome editing approaches. Single base mutations, substitutions, short deletions, and insertions account for 80% of genetic diseases. Gene correction in patient-derived pluripotent stem cells through programmable nucleases may present new healing options for patients in various congenital and acquired disorders (Pan *et al.*, 2013).

CRISPR-Cas9 mediated genome alteration can revolutionize the development of transgenic models by expanding biological research into the generation of novel genetically stable animal model organisms. Fertilized zygotes can be directly injected with Cas9 protein and transcribed sgRNA to acquire genetic modification at one and/or more loci in models, for example, rodents and monkeys. By evading the classic embryonic stem cell targeting stage in creating mutant lines, the production time for transgenic mice and rats might be reduced to only a few weeks from more than a year. Such progress will enable worthwhile and large-scale *in vivo* mutagenesis studies in models such as rodents can be pooled with explicit modification to avoid off-target mutagenesis. Furthermore, Cas9 can be exploited to directly alter somatic tissue, precluding the requirement for embryonic modification and permitting curative use for gene therapy. For generating cellular models, the targets cell can be simply transfected with a plasmid harboring Cas9 and aptly constructed sgRNA. Moreover, the composite potential of Cas9 proposes a promising way for analyzing widespread polygenic human diseases like diabetes, cardiovascular disease, schizophrenia, and autism (Bowdin, Ray, Cohn, & Meyn, 2014).

PERSONALIZED MEDICINE

Personalized medicine is referred to medical treatment of the individual characteristics using genetic information of the patient, not only to improve diagnostic and disease treatment but also a potential role in early detection and treatment of the disease. The advancement of emerging technologies like DNA sequencing, proteomics, wireless health monitoring devices, and imaging protocols have a different response among individuals regarding mechanisms and effects during disease processes. The different individuals respond with different effects during disease treatment due to unique molecular, environmental exposure, physiological, and behavior levels. Personalized medicine is the only alternative way to overcome disease-related problems by using an individual genetic profile. Individuals can get benefits from personalized medicine by sharing their genomic information. Personalized medicine has a tremendous role in health care of mankind. It is designed according to the individual genetic, proteomics, disease diagnostic and treatment. By using genetic information, the individual can prevent side effects are any disease-related problems. Genomic information of the individual helps the health scientist to choose appropriate therapy for disease treatment (Jain, 2002).

Personalized medicine leads to the development of pharmacogenomics. In pharmacogenomics, the effects of genes against particular drugs can be studied. This branch helps the scientist study the effect of the medicine on responder and non-responder, drug dosages, and adverse effects on the individual. Genes and proteins were used as a biomarker in personalized medicine to dragonize the disease. Moreover, the genomics of the individual also predicts the individual variability to a specific drug. The individual response helps the pharmacist to prescribe a suitable drug against a specific disease. The individual's genomics has a landmark role in drug designing and medical treatment (Hamburg & Collins, 2010).

The personalized medicine patients can be categorized based on genetic polymorphism of cytochrome p450 and enzyme polymorphism. The genomic information of patients helps to predict gene-drug interaction. The optimized drug efficacy and adverse effects can be studied through gene-drug interaction. The gene associated with metabolism and the immune system can be studied through gene-drug interaction. Gene-based drug targeting can also be studied through individual genomics. The molecular mechanism for a specific disease treatment was also studied through a gene-based drug targeting approach. The prediction and disease diagnostic were accurately studied through genomics information. The asthmatic patients were treated with the inhalation of the corticosteroids and β2-adrenergic. The patient with asthma showed less responsiveness and resistance to these treatments. The asthma patient's genomic helps the scientist study β2-adrenergic receptors for better treatment (Jain & Jain, 2009).

The genomic of personalized medicine have a tremendous benefit for drug designing rather than trial and error method. The life-threatening adverse effects can be minimized through personalized medicine. The cost of the clinical trial of a drug on animal models and human trials can be reduced through the genomics of individual personalized medicine. The failure of the drugs can be easily identified through genomic information. Moreover, the favorable response of the drugs can also study through genomics. The efficacy of the drugs can be improved with the aid of the genomics of personalized medicine (Liao & Tsai, 2013).

EPIGENOME

Epigenomic assessment includes examining alterations that don't include DNA nucleotide bases yet rather see things like DNA packaging and chemical alterations. These modifications of non-nucleotide have implications of regulating gene, which means whether gene expresses or not transcribed. Epigenomics all in all envelops histone alterations, limited changes to chromatin structure, exercises of non-coding RNAs, and DNA methylation. Methylation is a chemical modification of DNA that is currently mostly used in diagnostics as an epigenomic modification. DNA hypermethylation keeps a gene from being transcribed, bringing about the silencing of gene, while gene hypomethylation that is considered to be silenced outcomes in gene activation. DNA methylation is a normal process in order to regulate genes in a cell and is vital for controlling the cell climate explicit to a specific cell type with a specific task to take care of. Notwithstanding, gene hyper-or hypomethylation that shouldn't be likely clarifies a great deal of the components behind the regulation of gene variations from the norm noted in numerous problems, for example, Cancer. Epigenomics, as most approaches of genomic studies, assumes a double part in that it tends to be utilized to gather the information that permits us to comprehend a phenomenon of Nature more readily and subsequently better comprehends medical issues identified with that biological mechanism. It can likewise be used in a more, engaging way for hereditary testing (Bernstein, Meissner, & Lander, 2007).

Epigenomic approaches could be utilized to more readily comprehend fluctuation in patient results, symptomatology experienced by patients, *etc*. While the likely utility of epigenomic approaches is extraordinary, one must think about a portion of the limits and prerequisites for planning and understanding an examination. One huge issue is that the cell/tissue to be used is vital. Studies that research nucleotide organization of DNA (*e.g.*, an examination that utilizes polymorphisms to explore the genome) have the bit of leeway that the cell/tissue type from which the DNA was removed doesn't make any difference. This is not the situation for an epigenomic study because the methylation status and packaging of DNA will be explicit to the cell/tissue type utilized. In this way, when planning or assessing the writing on epigenomic discoveries, one should observe whether the cell/tissue from which the DNA was removed for the examination fitted for phenotype under scrutiny. Furthermore, dissimilar to polymorphism-based DNA examinations, the epigenomic status of the DNA is dynamic and can change over the long haul and in light of endogenous and exogenous conditions. Thus, one necessity to offer samples to whether tests were reliably gathered, concerned possible transient and earth-instigated changes. On a positive note, one bit of leeway of utilizing epigenomics approach is that the template of interest, DNA, is steadier than RNA; in this manner, it is more agreeable to test assortment by patients in the home and test assortment in the field, and there's no requirement for test obsession or adjustment (Murrell, Rakyan, & Beck, 2005).

CONCLUSION

This chapter presents a detailed description of possible health enhancement, Genome-wide association studies (GWAS), and next-generation sequencing (NGS). Moreover, the impacts of genomics on future health, genome editing, personalized medicine, and epigenome are also studied.

REFERENCES

Behjati, S., Tarpey, P.S. (2013). What is next generation sequencing? *Arch. Dis. Child. Educ. Pract. Ed., 98*(6), 236-238.

Bernstein, B.E., Meissner, A., Lander, E.S. (2007). The mammalian epigenome. *Cell, 128*(4), 669-681.

Bowdin, S., Ray, P.N., Cohn, R.D., Meyn, M.S. (2014). The genome clinic: a multidisciplinary approach to assessing the opportunities and challenges of integrating genomic analysis into clinical care. *Hum. Mutat., 35*(5), 513-519.

Bush, W.S., Moore, J.H. (2012). Chapter 11: Genome-wide association studies. *PLOS Comput. Biol., 8*(12)e1002822

Carbery, I.D., Ji, D., Harrington, A., Brown, V., Weinstein, E.J., Liaw, L., Cui, X. (2010). Targeted genome modification in mice using zinc-finger nucleases. *Genetics, 186*(2), 451-459.

Carroll, D. (2014). Genome engineering with targetable nucleases. *Annu. Rev. Biochem., 83*, 409-439.

Warman Chardon, J., Díaz-Manera, J., Tasca, G., Bönnemann, C.G., Gómez-Andrés, D., Heerschap, A., Mercuri, E., Muntoni, F., Pichiecchio, A., Ricci, E., Walter, M.C., Hanna, M., Jungbluth, H., Morrow, J.M., Fernández-Torrón, R., Udd, B., Vissing, J., Yousry, T., Quijano-Roy, S., Straub, V., Carlier, R.Y. (2019). MYO-MRI diagnostic protocols in genetic myopathies. *Neuromuscul. Disord., 29*(11), 827-841.

Detsika, M.G., Goudevenou, K., Geurts, A.M., Gakiopoulou, H., Grapsa, E., Lianos, E.A. (2021). Generation of a novel decay accelerating factor (DAF) knock-out rat model using clustered regularly-interspaced short palindromic repeats, (CRISPR)/associated protein 9 (Cas9), genome editing. *Transgenic Res., 30*(1), 11-21.

Feero, W.G., Wicklund, C., Veenstra, D.L. (2013). The economics of genomic medicine: insights from the IOM Roundtable on Translating Genomic-Based Research for Health. *JAMA, 309*(12), 1235-1236.

Gansauge, M-T., Gerber, T., Glocke, I., Korlević, P., Lippik, L., Nagel, S., Riehl, L.M., Schmidt, A., Meyer, M. (2017). Single-stranded DNA library preparation from highly degraded DNA using T4 DNA ligase. *Nucleic Acids Res., 45*(10), e79-e79.

Ghosh, D., Kumar, A., Sinha, N. (2021). Targeted genome editing: a new era in molecular biology.*Advances in Animal Genomics*. Elsevier.

Hamburg, M.A., Collins, F.S. (2010). The path to personalized medicine. *N. Engl. J. Med., 363*(4), 301-304.

Hu, J.H., Davis, K.M., Liu, D.R. (2016). Chemical biology approaches to genome editing: understanding, controlling, and delivering programmable nucleases. *Cell Chem. Biol., 23*(1), 57-73.

Jain, K.K. (2002). Personalized medicine. *Curr. Opin. Mol. Ther., 4*(6), 548-558.

Jain, K.K., Jain, K. (2009). *Textbook of Personalized Medicine..* Springer.

Kessel, M., Frank, F. (2007). A better prescription for drug-development financing. *Nat. Biotechnol., 25*(8), 859-866.

Ku, C.S., Loy, E.Y., Pawitan, Y., Chia, K.S. (2010). The pursuit of genome-wide association studies: where are we now? *J. Hum. Genet., 55*(4), 195-206.

Lee, J., Chung, J.H., Kim, H.M., Kim, D.W., Kim, H. (2016). Designed nucleases for targeted genome editing. *Plant Biotechnol. J., 14*(2), 448-462.

Liao, W.-L., Tsai, F-J. (2013). Personalized medicine: a paradigm shift in healthcare. *Biomedicine (Taipei), 3*(2), 66-72.

Mardis, E.R. (2008). Next-generation DNA sequencing methods. *Annu. Rev. Genomics Hum. Genet., 9*, 387-402.

Mattick, J.S., Dziadek, M.A., Terrill, B.N., Kaplan, W., Spigelman, A.D., Bowling, F.G., Dinger, M.E. (2014). The impact of genomics on the future of medicine and health. *Med. J. Aust., 201*(1), 17-20.

McBride, C.M., Bowen, D., Brody, L.C., Condit, C.M., Croyle, R.T., Gwinn, M., Khoury, M.J., Koehly, L.M., Korf, B.R., Marteau, T.M., McLeroy, K., Patrick, K., Valente, T.W. (2010). Future health applications of genomics: priorities for communication, behavioral, and social sciences research. *Am. J. Prev. Med., 38*(5), 556-565.

Mehta, R.K., Singh, J. (1999). Bridge-overlap-extension PCR method for constructing chimeric genes. *Biotechniques, 26*(6), 1082-1086.

Murrell, A., Rakyan, V.K., Beck, S. (2005). From genome to epigenome. *Hum. Mol. Genet., 14*(Spec No 1) (Suppl. 1), R3-R10.
http://dx.doi.org/10.1093/hmg/ddi110 PMID: 15809270

Naeimi Kararoudi, M., Hejazi, S.S., Elmas, E., Hellström, M., Naeimi Kararoudi, M., Padma, A.M., Lee, D., Dolatshad, H. (2018). Clustered regularly interspaced short palindromic repeats/Cas9 gene editing technique in xenotransplantation. *Front. Immunol., 9*, 1711.

Pan, Y., Xiao, L., Li, A.S., Zhang, X., Sirois, P., Zhang, J., Li, K. (2013). Biological and biomedical applications of engineered nucleases. *Mol. Biotechnol., 55*(1), 54-62.
http://dx.doi.org/10.1007/s12033-012-9613-9 PMID: 23089945

Pareek, C.S., Smoczynski, R., Tretyn, A. (2011). Sequencing technologies and genome sequencing. *J. Appl. Genet., 52*(4), 413-435.

Scherer, S.W., Visscher, P.M. (2016). *Genome-wide Association Studies: From Polymorphism to Personalized Medicine..* Cambridge University Press.

Shao, K., Ding, W., Wang, F., Li, H., Ma, D., Wang, H. (2011). Emulsion PCR: a high efficient way of PCR amplification of random DNA libraries in aptamer selection. *PLoS One, 6*(9)e24910

Stadler, Z.K., Thom, P., Robson, M.E., Weitzel, J.N., Kauff, N.D., Hurley, K.E., Devlin, V., Gold, B., Klein, R.J., Offit, K. (2010). Genome-wide association studies of cancer. *J. Clin. Oncol., 28*(27), 4255-4267.

Sun, N., Zhao, H. (2013). Transcription activator-like effector nucleases (TALENs): a highly efficient and versatile tool for genome editing. *Biotechnol. Bioeng., 110*(7), 1811-1821.

Tam, V., Patel, N., Turcotte, M., Bossé, Y., Paré, G., Meyre, D. (2019). Benefits and limitations of genome-wide association studies. *Nat. Rev. Genet., 20*(8), 467-484.
http://dx.doi.org/10.1038/s41576-019-0127-1 PMID: 31068683

Yeo, S.K., Guan, J-L. (2017). Breast cancer: multiple subtypes within a tumor? *Trends Cancer, 3*(11), 753-760.
http://dx.doi.org/10.1016/j.trecan.2017.09.001 PMID: 29120751

Zwart, H. (2009). From utopia to science: challenges of personalised genomics information for health management and health enhancement. *Med. Stud., 1*(2), 155-166.

CHAPTER 10

Society and Ethics

Abstract: This chapter proposes a brief description of Ethical values of advancement in science; Moral perspective of animal cloning; Social responses towards scientific interventions; The Do's and DON'Ts.

Keywords: Animal cloning, Ethics in science, Social interventions.

ETHICAL VALUES OF ADVANCEMENT IN SCIENCE

Modern scientific endeavors encourage advanced and progressive development in all areas of research. In one way, science provides the solution to most vulnerable questions in medical, health, environment, industry, *etc.*, while on the other hand, many scientific discoveries need cautious follow-up due to harmful impact on the masses. Societal acceptability of scientific research requires sound ethical reflection. The complexity of modern sciences and the rapid progress in research and development requires that up-to-date information is available in order to make such reflection possible. Moreover, a society is mostly characterized by a rich plurality of ethical, legal, and cultural traditions. This plurality, all the more, calls for intensive exchange on normative aspects of science. Ethics in science has increasingly become an important issue in democratic societies, especially subsequent to the modern life sciences. After important discoveries in the field of biology, and ethical reflection of scientific work; its results have come into focus. Naturally, the complexity of bioethics understood as medical ethics is prominently represented. Furthermore, closely related topics, like environmental ethics, animal ethics which can be subsumed under a broader understanding of bioethics, also take up much space. Although ethics in science deals with different aspects of modern scientific endeavors, few areas are more red zone that need immediate consideration. This chapter covers applications of euthanasia, assisted reproductive techniques (surrogate parenting, semen sorting), personalization in medicine, and issues of transgender. Although these scientific methods have been developed for therapeutic and beneficial purposes, as no ethical framework is defined, these methods are exploited in modern society. There is a need to develop solid ethical lines to effectively implement modern sciences in our society (Shrader-Frechette, 1994).

Maryam Javed, Asif Nadeem & Faiz-ul Hassan

MORAL PERSPECTIVE OF ANIMAL CLONING

Among various therapeutic techniques, animal cloning, has been a controversial topic that is still struggling to get a place in successful commercial therapeutics (Fig **10.1**). Animal cloning is a method that allows creating animals of desirable identical traits. Somatic cells of the donor are isolated and enucleated. The nucleus is inserted into the enucleated zygote, and then an electrical signal is used to trigger the cell division. The newly formed zygote is implanted into a surrogate mother, and a clone is born after completing the gestational period. Due to the low efficiency of the method, there is a loss of embryos at many stages that raise questions over the ethical perspective of this method. Generating similar individuals by this experiment is another question in the minds of many about the outcomes of clones and their societal status. Are these individuals? Are they claiming rights equally? OR are we playing with God by creating similar individuals? Contrary to this, supporters of this method defend the use of animal cloning in saving endangered animals, enhancing the phenotypic potential of livestock, creating living bioreactors for many biopharmaceuticals, *etc.*

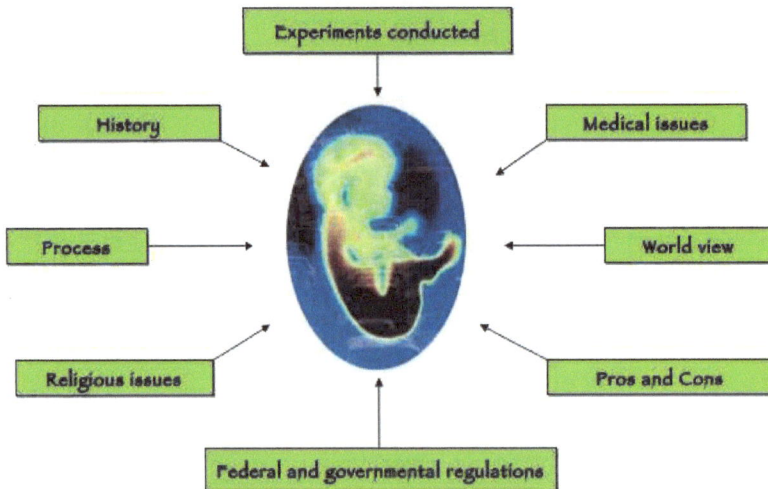

Fig. (10.1). Ethical Concerns for Cloning.

SOCIAL RESPONSES TOWARDS SCIENTIFIC INTERVENTIONS

Science and society have been at war since the beginning. Although science is all about the benefits of humankind and society, many scientific methods are still not supported by society and scientists. All those areas of science, directly and indirectly, affect human existence, moral/cultural values, and religion comes under the umbrella of bioethics. Different modern civilizations have already defined the rules and remits for the scientific community as use of the living model for experimentation, cloning, stem cells, surrogating mothers, assisted reproductive techniques, GMOs, *etc.* some of them are discussed here;

Use of Living Models For Experimentation

Animal models are used in many different experiments such as drug trials, preventive medicines, surgical techniques, anatomical studies, collecting tissues, blood, extracting hairs, studying the effect of pollutants, drugs, *etc.* The use of living models in science was first raised in 1822, and the first time British Parliament defined law to protect living research models. Since then, the concept of animal protection has consisted of the idea of "3R".

Replacement: It is to find suitable alternatives for the living model. The first level replaces the higher model organisms with the lower ones. Instead of using primates, rodents are used, or computer simulations or non-living tissues should be proffered instead of using living organisms.

Reduction: The number of animals required to experiment should be reduced to n minimal acceptable number. This can be avoided by cutting the unnecessary replication of the experiment without disturbing the statistical significance.

Refinement: Techniques and experimental procedures used on animals should be minimally invasive and should not be prone to any additional stress in the animals. Animal confinement should not be prolonged. Experiments dealing with restrictive dietary requirements should also not be too long. If animals are not able to continue their lives after experiments, unpainful Euthanasia should be practiced.

Stem Cells

These are the primitive, rare, undifferentiated cells in the body with tremendous potential in regenerative medicine. Stem cells can be collected from early embryos. This practice has raised many concerns over the loss of embryos for mere collection of this cell line. This has caused a significant increase in the annual abortion rate in many of the developing countries. Ethical remits are not allowing the use of fetus and embryo for the extraction of stem cells. They are encouraging the use of adult stem cells and induced pluripotent stem cells as an alternative and safe option.

Surrogate Mothers

This method was established as a therapeutic solution in assisted reproductive technique but has been highly exploited, especially in developing and underdeveloped countries. People are now compelling young girls to be paid surrogates without accounting for the health hazards related to this. Ethical laws are now in practice for surrogating and should be followed strictly to avoid inconveniences.

GMOs

Genetically modified organisms capture the market at a remarkable pace and have replaced organic food on food shelves. Although the advantages of GMOs are remarkable, we cannot neglect the ethical concerns of the community for this kind of food. Food sources of this origin have been declared unsafe in many societies for health and the environment. These can affect the genetic diversity of an ecosystem (Rhodes, Closson, Paparini, Guise, & Strathdee, 2016).

THE *DO'S* AND *DON'TS*

Ethical roles and remits firstly came into the discussion in the early 1900's, where researchers and scientists raised questions over the practical efficacy of many therapeutic techniques compared to their invasive and hazardous impact on the general public. The National Commission for the Protection of Human Subjects of Biomedical and Behavioral Research was initially established in 1974 to define the basic ethical principles in different disciplines of science, biology, medicine, and research. According to their regulations, no such techniques can be allowed for masses where the potential benefits underweight the risks associated with this research. These regulations cover the majority of the concerns as a matter of living animal models, use of humans in clinical trials, the matter of patient's consent, financial support, the matter of intellectual property rights, plagiarism, research duplication, research paper authorship claims, deception, protection of the subjects included in any research (Bach & Meigen, 1999);(Dror, 2005);(Orive, Emerich, & De Vos, 2014);(Lazarides, Georgiadis, & Papanas, 2020).

CONCLUSION

This chapter presents a detailed description of ethical values of advancement in science, ethical concerns for cloning, moral perspective of animal cloning, and responses towards scientific interventions. Moreover, the use of living models for experimentation is discussed.

REFERENCES

Fiester, A. (2005). Ethical issues in animal cloning. *Perspect. Biol. Med., 48*(3), 328-343.
http://dx.doi.org/10.1353/pbm.2005.0072 PMID: 16085991

Bach, M., Meigen, T. (1999). Do's and don'ts in Fourier analysis of steady-state potentials. *Doc. Ophthalmol., 99*(1), 69-82.
http://dx.doi.org/10.1023/A:1002648202420 PMID: 10947010

Dror, I. (2005). Experts and technology: Do's & Don'ts. *Biometric Technology Today, 13*(9), 7-9.
http://dx.doi.org/10.1016/S0969-4765(05)70429-X

Lazarides, M.K., Georgiadis, G.S., Papanas, N. (2020). Do's and don'ts for a good reviewer of scientific papers: a beginner's brief decalogue. *Int. J. Low. Extrem. Wounds, 19*(3), 227-229.
http://dx.doi.org/10.1177/1534734620924349 PMID: 32525721

Orive, G., Emerich, D., De Vos, P. (2014). Encapsulate this: the do's and don'ts. *Nat. Med., 20*(3), 233-233.
http://dx.doi.org/10.1038/nm.3486 PMID: 24603789

Rhodes, T., Closson, E.F., Paparini, S., Guise, A., Strathdee, S. (2016). Towards "evidence-making intervention" approaches in the social science of implementation science: The making of methadone in East Africa. *Int. J. Drug Policy, 30*, 17-26.
http://dx.doi.org/10.1016/j.drugpo.2016.01.002 PMID: 26905934

Shrader-Frechette, K.S. (1994). *Ethics of Scientific Research.*. Rowman & Littlefield.

SUBJECT INDEX

A

Acetylation process 85, 86
Acid (RNA), Ribonucleic 1, 7, 8, 9, 25, 26,
    28, 29, 34, 109, 111, 112, 114, 132, 163
Activity 14, 38, 69, 166
    exonuclease 14
    methylase 69
    much-needed 38
    protein's 166
Acute immune rejection 168
AFLP technology 44
Agarose gel electrophoresis 43
Amplified fragment length polymorphism 44
Angiotensin-converting enzyme (ACE) 119
Antibiotics penicillin 75
Apoptosis 85, 86
Artificial intelligence 123
Asthma 118, 149, 152, 170
ATP 43, 71, 81, 82, 156
    dependent exchange 82
    hydrolysis 81
    sulfurylase 156
Autoradiography 43, 71

B

Bacillus 88, 89
    anthracis 89
    cereus 88
Bacteria 65, 67, 75, 88, 121, 122, 127, 164
    artificial chromosome 65
    disease-causing 75, 88
    plant's pathogen Xanthomonas 164
BEVS expression system 67
Big data analysis 121, 122, 127
Bioinformatics 40, 53, 86, 117
    methods 53
    software 40
    software for sequence data analysis 117
    technology 86
Breeding programs 62

Bronchial hyper responsiveness (BHR) 152
Brownian movement 42

C

CADD methods 119
Cancer 78, 84, 86, 88, 95, 96, 100, 137, 138,
    148, 152, 157, 171
    breast 138, 152
    genome atlas 137
    growth 157
    lung 138
    prevention 95, 96, 100
Cancerous tissue 126
Causal role (CR) 111, 112
Cell(s) 2, 161
    cytoplasm 8
    repair mechanism 161
Cellular differentiation process 78
Chargaff's rules 2
Chemotherapy 157
Chinese hamster ovary cells 168
Chromatin dynamics 85
Chromatogram 48, 49
Chromatography 59
    gas 59
Chromosomal inheritances 78, 82
Chromosome banding methods 95
Computer-aided drug designing 104, 119, 127
Crick's version 26
Cryo-electron microscopy reconstructions 166
Cyanobacteria 8, 108
Cystic fibrosis 73
Cytochrome 45, 115, 118, 170
    mitochondrial 45
Cytogenetic procedures 95
Cytosine nucleotides 84
    altered 84
    methylated 84

Maryam Javed, Asif Nadeem & Faiz-ul Hassan
All rights reserved-© 2021 Bentham Science Publishers

www.ingramcontent.com/pod-product-compliance
Lightning Source LLC
Chambersburg PA
CBHW041701210326
41598CB00007B/485